FOOD SCIENCE AND TECHNOLOGY SERIES

FOOD SAFETY, FRESH PRODUCE AND FDA OVERSIGHT

FOOD SCIENCE AND TECHNOLOGY SERIES

Food Science and Technology: New Research
Lorenzo V. Greco and Marco N. Bruno (Editors)
2008. ISBN: 978-1-60456-715-1

Food Science and Technology: New Research
Lorenzo V. Greco and Marco N. Bruno (Editors)
2008. ISBN: 978-1-61668-106-7 (Online Book)

The Price of Food
Meredith N. Fisher (Editor)
2009. ISBN: 978-1-60692-440-2

Food Processing and Engineering Topics
Maria Elena Sosa-Morales and Jorge F. Velez-Ruiz (Editors)
2009. ISBN: 978-1-60741-788-0

Traditional Chinese Foods: Production and Research Progress
Li Zaigui and Tan Hongzhuo
2009. ISBN 978-1-60692-902-5

Traditional Chinese Foods: Production and Research Progress
Li Zaigui and Tan Hongzhuo
2009. ISBN: 978-1-61668-277-4 (Online Book)

Food Science Research and Technology
Isaak Hülsen and Egon Ohnesorge (Editors)
2009. ISBN 978-1-60741-848-1

Food Hydrocolloids: Characteristics, Properties and Structures
Clarence S. Hollingworth (Editor)
2010 ISBN: 978-1-60876-222-4

The Olive: Its Processing and Waste Management
José S. Torrecilla
2010. ISBN: 978-1-60876-719-9

Food Safety, Fresh Produce and FDA Oversight
Russell P. Barton (Editor)
2010. ISBN: 978-1-60692-788-5

FOOD SCIENCE AND TECHNOLOGY SERIES

FOOD SAFETY, FRESH PRODUCE AND FDA OVERSIGHT

RUSSELL P. BARTON
EDITOR

Nova Science Publishers, Inc.
New York

LIBRARY OF CONGRESS CATALOGING-IN-PUBLICATION DATA

Food safety, fresh produce, and FDA oversight / editor, Russell P. Barton.
p. cm.
Includes index.
ISBN 978-1-60692-788-5 (hardcover)
1. Food adulteration and inspection--Government policy--United States. 2. Farm produce--Security measures--United States. 3. Food contamination--United States. 4. United States. Food and Drug Administration. I. Barton, Russell P.
TX531.F56825 2009
353.9'970973--dc22

2009032065

Published by Nova Science Publishers, Inc. ✢ New York

TABLE OF CONTENTS

PREFACE

In recent years, both domestic and imported produce have been linked to reported outbreaks of foodborne illness. Contamination in produce is of particular concern because produce is often consumed raw. The Food and Drug Administration (FDA) has primary responsibility for ensuring the safety of both domestic and imported fresh produce. The authors were asked to examine (1) the resources FDA has spent on fresh produce safety and how it has allocated those resources, (2) the effectiveness of FDA's actions to oversee fresh produce safety, and (3) the extent to which FDA's planned actions to enhance fresh produce oversight address identified challenges. For this review, the authors analyzed FDA spending data and estimates and FDA activities data, reviewed FDA plans, and interviewed FDA officials and others.

Chapter 1 - In recent years, both domestic and imported produce have been linked to reported outbreaks of foodborne illness. Contamination in produce is of particular concern because produce is often consumed raw. The Food and Drug Administration (FDA) has primary responsibility for ensuring the safety of both domestic and imported fresh produce. GAO was asked to examine (1) the resources FDA has spent on fresh produce safety and how it has allocated those resources, (2) the effectiveness of FDA's actions to oversee fresh produce safety, and (3) the extent to which FDA's planned actions to enhance fresh produce oversight address identified challenges. For this review, GAO analyzed FDA spending data and estimates and FDA activities data, reviewed FDA plans, and interviewed FDA officials and others.

Chapter 2 - Numerous federal, state, and local agencies share responsibilities for regulating the safety of the U.S. food supply, which many experts say is among the safest in the world. Nevertheless, critics view this system as lacking the organization and resources to adequately combat foodborne illness — as evidenced by a series of widely publicized food safety problems, including concerns about adulterated food and food ingredient imports, and illnesses linked to various types of fresh produce and to meat and poultry products. Numerous bills addressing various aspects of food safety have been introduced into the first session the 110th Congress, including proposals to reorganize oversight authorities, increase funding, tighten regulation of imported foods, and establish new authorities for enforcement, recall, and notification regarding adulterated foods, among others. Portions of several proposals have become law; more sweeping changes in the federal food safety system are still under consideration in 2008.

Chapter 3 - The Food and Drug Administration Amendments Act of 2007 (FDAAA; P.L.110-85), while primarily concerned with drug and device regulation and their user fees, also contains several provisions on food safety. As enacted, the law requires the Secretary of

Health and Human Services to establish, for pet foods, processing and ingredient standards, updated labeling requirements, and an early warning and surveillance system to identify adulteration and associated outbreaks of illness. The Secretary is to work with states to improve the safety of produce and strengthen state food safety programs. The Act requires the Secretary to create a registry for reporting of foods (including human and animal foods) with safety problems, which will help identify the supply chain of the reportable food. Alerts are to be issued for such foods, with records maintained and available for inspection. Additional provisions require attention to aquaculture and seafood inspection, environmental risks associated with genetically engineered seafood products, imported foods, pesticide monitoring, and ginseng dietary supplements.

On September 27, 2007, the President signed the Food and Drug Administration Amendments Act of 2007 (FDAAA; P.L. 110-85, H.R. 3580) into law. The Act primarily addresses issues such as regulation and user fees for prescription drugs and medical devices, clinical trials, pediatric research incentives, and conflicts of interest.[1] However, the Act also contains several provisions on food safety that address recent concerns about contaminated pet food and tracking adulterated foods. This chapter reviews the food safety provisions of the statute, their history, and certain issues surrounding their passage. It will not be updated.

Chapter 4 - The Food and Drug Administration (FDA) is the agency within the Department of Health and Human Services (HHS) that regulates human and animal drugs, medical devices, biologics, and most foods. This chapter describes FDA, surveys agency-related issues Congress faces, and cites CRS reports where readers can find more information.

Chapter 5 - FDA is the Federal agency that regulates almost everything we eat except for meat, poultry, and processed egg products, which are regulated by our partners at the United States Department of Agriculture (USDA). FDA is committed to ensuring that America's food supply continues to be among the safest in the world.

There is no question that the *Salmonella* Saintpaul outbreak investigation has been one of the most complex investigations in recent memory. I assure you that FDA is committed to working with all our food safety partners to examine ways to remove or mitigate some of the complicating factors to expedite tracebacks. In my testimony, I will discuss some of the factors that made this investigation so complex. I will also describe some of the challenges we face both in preventing fresh produce from becoming contaminated in the first place and in investigating outbreaks associated with fresh produce. I will also discuss some of the specific measures FDA is taking to enhance the safety of fresh produce and other foods to prevent future outbreaks and to improve traceability when an outbreak occurs.

Food can become contaminated at many different steps – on the farm, in processing or distribution facilities, during transit, at retail and food service establishments, and in the home. In recent years, we have done a great deal to prevent both intentional and unintentional contamination of food at each of these steps. FDA has worked with other Federal, state, local, tribal, and foreign counterpart food safety agencies, as well as with law enforcement and intelligence-gathering agencies, and with industry, consumer groups, and academia to significantly strengthen the nation's food safety and food defense system across the entire distribution chain.

Chapter 6 - FDA is committed to ensuring that America's food supply continues to be among the safest in the world. In recent years, we have done a great deal to protect the food supply from both unintentional and deliberate contamination. We have made significant progress in both, but the recent outbreaks of foodborne illness underscore the need to renew our focus on multidisciplinary and integrated food safety strategies.

The Commissioner of Food and Drugs, Dr. Andrew von Eschenbach, recently appointed me to the newly created position of Assistant Commissioner for Food Protection. My first priority in this position is to develop a new strategy for food safety and food defense that will address changes in the global food safety and food defense system, identify our most critical needs, and serve as a framework to help us address the challenges we face. The goal is to ensure a comprehensive and robust food safety and food defense program that is tailored to meet the risks posed by the types of foods we regulate and that focuses on prevention, ensures compliance with preventive controls, and rapidly responds when contaminated food or feed is detected, or when there is harm to humans or animals.

The number of illnesses associated with fresh produce is a continuing concern of the Agency, and we have worked on a number of initiatives to reduce the presence of pathogens in these foods. Fresh produce presents special challenges. For example, consumption of produce, particularly "ready-to-eat" products, has increased dramatically during the past decade. This is a positive development from a nutrition perspective, but also a new dynamic that challenges our food safety efforts.

Because most produce is grown in an outdoor environment, it is vulnerable to contamination from pathogens that may be present in the soil, in agricultural or processing water, and in manure used as fertilizer, or due to the presence of animals in or near fields or packing areas. Produce also is vulnerable to contamination due to inadequate worker health and hygiene protections, environmental conditions, inadequate production safeguards, and inadequate sanitation of equipment and facilities. The fact that produce is often consumed raw or with only minimal processing, without any type of intervention that would reduce or eliminate pathogens prior to consumption, contributes to its potential as a source of foodborne illness. Consequently, addressing the way fresh produce is grown, harvested, and moved from field to fork is crucial to minimizing the risk of microbial contamination.

Chapter 7 - Lonnie J. King addresses the Subcommittee on CDC's activities related to the prevention of foodborne disease and CDC's role in the response to the current outbreak of *Salmonella* Saintpaul infections associated with fresh produce. First, he offers his sympathies to all the families who have been adversely affected by this outbreak. Second, he understand the frustration of many in the food producing and serving industries, who work very hard to produce and serve safe produce. This investigation has been especially difficult and prolonged. They have faced many challenges with this particular foodborne outbreak. He will discuss these challenges in more detail after describing the CDC's response to the *Salmonella* Saintpaul outbreak.

In: Food Safety, Fresh Produce and FDA Oversight
Editor: Russell P. Barton
ISBN: 978-1-60692-788-5

Chapter 1

FOOD SAFETY: IMPROVEMENTS NEEDED IN FDA OVERSIGHT OF FRESH PRODUCE

United States Government Accountability Office

Why GAO Did This Study

In recent years, both domestic and imported produce have been linked to reported outbreaks of foodborne illness. Contamination in produce is of particular concern because produce is often consumed raw. The Food and Drug Administration (FDA) has primary responsibility for ensuring the safety of both domestic and imported fresh produce. GAO was asked to examine (1) the resources FDA has spent on fresh produce safety and how it has allocated those resources, (2) the effectiveness of FDA's actions to oversee fresh produce safety, and (3) the extent to which FDA's planned actions to enhance fresh produce oversight address identified challenges. For this review, GAO analyzed FDA spending data and estimates and FDA activities data, reviewed FDA plans, and interviewed FDA officials and others.

What GAO Recommends

GAO recommends, among other things, that the Commissioner of FDA update its guidance on good agricultural practices and its regulations on current good manufacturing practice for food, and seek explicit authority from the Congress to adopt preventive controls for high-risk foods and authority for enhanced access to records.

FDA agreed with most of GAO's recommendations but believed that it had sought authority from the Congress. FDA should continue to take steps to obtain these authorities so that it can conduct its oversight responsibilities.

To view the full product, including the scope and methodology, click on GAO-08-1047.

For more information, contact Lisa Shames at (202) 512-3841 or shamesl@gao.gov.

What GAO Found

While FDA has considered fresh produce safety a priority for many years, resource constraints and other work—including counterterrorism efforts and unplanned events such as foodborne illness outbreaks—have caused FDA to delay key produce safety activities. FDA has no formal program devoted exclusively to fresh produce and has not consistently and reliably tracked its fresh produce spending. Based on FDA estimates, FDA spent at least $20 million and 130 staff years on fresh produce in fiscal year 2007—or about 3 percent of its food safety dollars and 4 percent of its food safety staff years. In addition, FDA had few staff dedicated solely to fresh produce safety. Moreover, FDA acknowledged that it has not yet been able to conduct certain fresh produce work crucial to understanding the incidence of contamination of produce by pathogens such as *E. coli* O157:H7 or *Salmonella*, because it has lacked the resources to either fund its extramural research grant program or perform some critical research internally. Finally, FDA delayed issuing final fresh-cut produce guidance at least 6 years because it had to shift staff to counterterrorism and outbreak investigation work.

FDA has provided limited oversight of domestic and imported fresh produce. For example, while FDA has issued guidance for industry on recommended practices for reducing the risk of contamination during the processing of fresh-cut produce, it has not issued regulations requiring firms to take action to prevent contamination, even though some industry groups would like it to do so. FDA's intervention efforts have also been limited. Specifically, domestic fresh produce firms were inspected infrequently. Furthermore, FDA examined less than 1 percent of the 7.6 million fresh produce lines imported from fiscal years 2002 through 2007. Finally, FDA has improved some elements of its emergency response by, for example, partnering with California on outbreak investigations. However, it faces challenges in tracing an outbreak involving fresh produce back to its source because produce is highly perishable and may no longer be available for testing. Also, when product is available, it may be unlabeled or mixed in packages containing products from multiple sources.

FDA has proposed changes through its *Food Protection Plan* that could significantly enhance its fresh produce oversight. However, the agency is still in the planning stages for several enhancements and has not provided specific information on strategies and resources, making it difficult to assess the likelihood of success. To help prevent contamination, FDA plans to update its existing guidance on good agricultural practices and regulations on current good manufacturing practice for food, and has identified a need for explicit authority to issue preventive safety regulations for high-risk foods and enhanced access to records. To enhance intervention efforts, FDA plans to use more rigorous risk-based criteria to target domestic firm inspections and is testing a new import screening software tool. To improve response efforts, FDA is examining best practices for tracing contaminated foods to their source.

September 26, 2008

The Honorable Edward M. Kennedy
Chairman
Committee on Health, Education, Labor, and Pensions
United States Senate

The Honorable Barbara Boxer
United States Senate

In recent years, there has been an increase in reported outbreaks of foodborne illness associated with both domestic and imported produce. Most recently, an outbreak of *Salmonella* linked to fresh produce, which sickened at least 1,440 people in 43 states, the District of Columbia, and Canada, has become the largest foodborne illness outbreak reported in the last 10 years. According to the Centers for Disease Control and Prevention (CDC), fresh produce was linked to 7 percent of all outbreaks that had been traced to a specific food source and 14 percent of the associated illnesses from 1998 to 2004. In addition to harming human health, outbreaks of foodborne illness can undermine consumer confidence in the safety of the nation's food supply and have serious economic consequences. The 2006 outbreak of *E. coli* O157:H7 linked to bagged spinach, for example, resulted in 205 confirmed illnesses, 3 deaths, and an estimated $100 million loss to industry. The importance of this issue is growing because the consumption of fresh produce has increased as both health experts and the U.S. government have encouraged Americans to eat fruits and vegetables as part of a healthy diet. According to the U.S. Department of Agriculture (USDA), the average American annually consumed 13 pounds more fresh fruit and 50 pounds more fresh vegetables from 2003 through 2005 than from 1983 through 1985, an increase of about 14 percent and 41 percent, respectively. Also, more people have turned to the convenience of fresh-cut produce, such as bagged salads and cut fruit, and significant increases in imported produce have made a greater variety and volume of fresh produce available year round.

Produce is particularly vulnerable to contamination with pathogens (microorganisms that can cause disease) because it is grown in a natural environment. Moreover, it is often consumed raw, without cooking or other treatment that would reduce, control, or eliminate pathogens prior to consumption. Processing produce into fresh-cut products, the fastest growing segment of the fresh produce market, may increase the risk of microbial contamination and growth by breaking the surface of the produce and allowing pathogens to enter the product. The contamination can then spread to other produce being processed at the same time. Fresh produce may also become contaminated after it is harvested and processed, such as during transportation, preparation, or storage.

The Food and Drug Administration (FDA) within the Department of Health and Human Services has primary responsibility for ensuring the safety of food for humans and animals. Specifically, FDA is responsible under the Federal Food, Drug, and Cosmetic Act for ensuring that domestic and imported human food (except meat, poultry, and processed egg products) and animal feed are safe, wholesome, and labeled properly. Under the Public Health Service Act, FDA has the authority to take measures to prevent the spread of disease. In addition, FDA may enter into arrangements with states to do inspections, share resources, or avoid duplication of efforts.

In January 2007, we added the federal oversight of food safety to our High-Risk Series, which is intended to raise the priority and visibility of government programs that are in need of broad-based transformation to achieve greater economy, efficiency, effectiveness, accountability, and sustainability.[1] In particular, we noted that federal spending for food safety oversight has not been commensurate with the volume of foods regulated by the agencies or consumed by the public. In November 2007, a report for FDA's science advisory board, *FDA Science and Mission at Risk*,[2] pointed out the erosion in FDA's science base. The report cited numerous management challenges that have contributed to FDA's inability to fulfill its mission that cannot be addressed with available resources, such as the lack of information sciences and infrastructure to support new science. That same month, a report by FDA, *Food Protection Plan: An Integrated Strategy for Protecting the Nation's Food Supply*,[3] recognized the need for several changes to ensure the safety of the nation's food supply, such as shifting efforts toward prevention, and identified new authorities needed to implement the new strategy. We have testified that FDA's *Food Protection* Plan—which presents a three-part framework of prevention, intervention, and response—proposes some positive first steps toward enhancing FDA's oversight of food safety. We also pointed out, however, that more information about strategies and the resources FDA needs to implement the plan would facilitate congressional oversight.[4] In particular, we noted that FDA's overall resource needs and timelines for fully implementing the plan are unclear. In June 2008, we testified that FDA had implemented few of our past recommendations to improve food safety oversight. Specifically, we had made a total of 34 food safety-related recommendations to FDA since 2004, and as of May 2008, FDA had implemented 7 of those recommendations. In commenting on a draft of this chapter, FDA stated that an update on the status of these recommendations may allow an additional 15 recommendations to be viewed as fully implemented. FDA also stated that two recommendations would require congressional action and one would require additional funding to implement. Based on our routine update on the status of open recommendations, we agree that one additional recommendation can be considered fully implemented; however, we disagree with FDA's assessment that the remaining recommendations should be considered fully implemented.

As requested, this chapter examines (1) the dollars and staff years FDA has spent on fresh produce safety and how FDA has allocated those resources, (2) the effectiveness of FDA's actions to oversee domestic and imported fresh produce safety, and (3) the actions FDA plans to take to enhance fresh produce oversight and the extent to which FDA's planned actions address identified challenges.

For this chapter, fresh produce means fruits and vegetables in their unpeeled, natural form, as well as fruits and vegetables that have been minimally processed (e.g., peeled, sliced, or chopped) before being packaged for use by the consumer or a retail establishment. It does not include frozen or canned fruits and vegetables or fruit and vegetable juices. To conduct this review, we visited produce farms and processing facilities in California's Salinas Valley, where we interviewed growers, processors, and industry representatives and observed an FDA inspection of a fresh-cut produce facility. We selected the Salinas Valley because it was the source of the 2006 *E. coli* 0157:117 outbreak linked to bagged spinach. We obtained and analyzed FDA data and estimates on food safety and fresh produce safety spending in both dollars and staff years and FDA data on fresh produce oversight activities. In analyzing FDA's food safety resources, we focused on fiscal years 2000 through 2007 to update a previous GAO report that detailed food safety spending through fiscal year 1999.[6] In analyzing FDA's fresh produce resources, we limited our work to fiscal years 2005 through 2007 because FDA believed it could only provide reliable estimates of fresh produce spending for these years. We assessed the reliability of the data used in this chapter and found it to be sufficiently reliable for the purposes used. We reviewed FDA plans, such as its November 2007 *Food Protection Plan,* for information on proposed changes that could enhance fresh produce oversight. To assess FDA's plans, we reviewed previous GAO reports on food safety and GAO guidance for assessing key elements in agencies' performance plans, including goals, strategies, and resources.[6] We interviewed FDA officials about fresh produce resources, oversight activities, and planned changes. We also interviewed former FDA officials, food safety experts, state food safety officials, industry representatives, and others to obtain their views on FDA's current oversight activities and planned actions. Appendix I contains a detailed discussion of the scope and methodology of our review.

We conducted this performance audit from June 2007 to September 2008 in accordance with generally accepted government auditing standards. Those standards require that we plan and perform the audit to obtain sufficient, appropriate evidence to provide a reasonable basis for our findings and conclusions based on our audit objectives. We believe that the evidence obtained provides a reasonable basis for our findings and conclusions based on our audit objectives.

RESULTS IN BRIEF

Although FDA has considered fresh produce as a priority over the past decade, resource constraints and other work—including counterterrorism efforts and unplanned events such as foodborne illness outbreaks—have caused FDA to delay key fresh produce safety initiatives. Because FDA has no formal program devoted exclusively to fresh produce, it draws dollars and staff years to fresh produce from its overall food safety program. FDA has not consistently and reliably tracked its spending on fresh produce, thus limiting its ability not only to identify its actual fresh produce spending but also to plan and manage this spending. Based on FDA's estimates, our analysis indicates that spending on fresh produce safety was at least $18 million in fiscal year 2006 and at least $20 million in 2007, or approximately 3 percent of total FDA food safety spending in each year. Similarly, our analysis shows that FDA spent at least 132 staff years on produce safety in fiscal year 2006 and 130 staff years in 2007, or about 4 percent of its total food safety staff years. Additionally, FDA had few staff dedicated solely to fresh produce safety. This low level of spending relative to total food safety spending is partly the result of resource constraints and other work that have delayed fresh produce efforts. For example, officials from FDA's Center for Food Safety and Applied Nutrition (CFSAN) told us that one of the center's priorities—issuing guidance for fresh-cut produce operations—was delayed at least 6 years because they had to divert staff with the needed expertise to address counterterrorism efforts and outbreaks of foodborne illness.

FDA's actions to oversee domestic and imported fresh produce safety have generally been limited. More specifically, within FDA's *Food Protection Plan* framework of prevention, intervention, and response, we found the following:

- *Prevention.* FDA's prevention efforts have been limited, in part because gaps in scientific knowledge have impeded its ability to fully integrate science and risk into its oversight of fresh produce safety. Moreover, FDA has issued some voluntary guidance for industry, including guidelines for minimizing contamination during field and fresh-cut operations. However, it has not issued regulations requiring firms to take action to prevent contamination, even though some industry groups would like it to do so.

- *Intervention.* Inspections of domestic firms that handle fresh produce have occurred infrequently. Our analysis of FDA data showed that the 2,002 domestic firms that underwent produce-related inspections were inspected twice, on average, from fiscal years 2000 through 2007. Problems were observed in 41 percent of these inspections, but most did not warrant further regulatory action, according to FDA. Therefore, the agency primarily relied on firms to take voluntary corrective action. We also found FDA provided minimal oversight of imported produce. Although FDA's oversight of imports relies heavily on screening products at the border, it examined less than 1 percent of the 7.6 million fresh produce import entry lines from fiscal years 2002 through 2007. Additionally, although FDA devoted more resources to import oversight, enabling it to conduct more import examinations in fiscal year 2007 than in fiscal year 2004, it has not been able to inspect a larger share of incoming fresh produce shipments.

- *Response.* FDA has improved some elements of its emergency response. Successes include improving response coordination through the creation of a new Office of Crisis Management, partnering with California on outbreak investigations, and developing a pilot program where recall notices include photographs of the recalled product. However, tracing an outbreak involving fresh produce back to its source remains challenging because produce is highly perishable and may no longer be available for testing. Also, when product is available, it may be unlabeled or mixed in packages with products from multiple sources.

FDA has proposed changes that could significantly enhance its fresh produce oversight and has reported some initial progress, but more information on strategies and resources is needed to enhance accountability and assess the likelihood of FDA's success. Specifically, through its *Food Protection Plan*, FDA proposed agency actions and identified authorities needed to better leverage its limited resources and strengthen its oversight of fresh produce, including these key actions:

- *Prevention.* FDA plans to help fill gaps in scientific knowledge and update its 1998 guidance to industry on good agricultural practices. Also, FDA has cited a need for explicit authority to issue regulations requiring preventive controls for high-risk foods, such as leafy greens, which could minimize the risk of contamination before such foods enter the market.

- *Intervention.* For domestic inspections, when deciding which domestic food facilities to inspect, FDA plans to increase the rigor of its risk-based criteria to focus on the firms of highest risk. For imports, FDA officials are also testing a new import screening software tool that uses information from a wider variety of sources to more effectively screen products at the border. Further, FDA has identified a need for explicit authority to accredit third parties to perform inspections, which could help FDA leverage its resources, but FDA does not envision using such a program for fresh produce until the agency has established or assessed the adequacy of clear preventive standards.

- *Response.* FDA plans to improve efforts to trace contaminated products back to their source by establishing an internal working group to examine industry best practices, using new laboratory equipment, and obtaining improved records access authority during food-related emergencies. FDA also plans to organize more federal-state teams, such as the FDA-California team, to respond to outbreaks. In addition, FDA has identified the need for the authority to issue mandatory recalls when voluntary recalls are not effective and has plans to improve how it communicates risk to the public during outbreaks, with help from an external advisory group.

While these efforts have the potential to enhance food safety oversight, FDA is still largely in the planning stages for these improvements and has not provided specific information on strategies and resources. Without this information, it is difficult to assess the likelihood of success.

We are making seven recommendations to FDA, including four recommendations to enhance its oversight of fresh produce safety, such as updating its good agricultural practices guidance, two recommendations to seek authority from the Congress to make explicit FDA's authority to adopt preventive controls for high-risk foods and to provide enhanced access to firm records during food-related emergencies, and one recommendation to provide specific information to the Congress and to the public on the strategies and resources for implementing the *Food Protection Plan*. In its written comments on a draft of our report that included comments from FDA, the Department of Health and Human Services generally agreed with the report's accuracy and conclusions and with most of the report's recommendations. While FDA agreed with the importance of having explicit authority to adopt preventive controls for high-risk foods and having enhanced access to firm records during food-related emergencies, the agency believes that it has already sought such authorities by outlining legislative needs in the *Food Protection Plan* and testifying on the plan before congressional committees. We do not view these actions as seeking authority. Rather, as FDA recognized, there is a need for the agency to partner with the Congress to make the necessary changes to transform the safety of the nation's food supply. FDA should move beyond outlining needs and continue to take steps to obtain these legislative authorities, such as by suggesting language that provides FDA the necessary statutory tools to help the agency conduct its oversight responsibilities. In addition, FDA provided technical comments that we have incorporated, as appropriate.

BACKGROUND

FDA has primary responsibility for ensuring the safety of a broad range of products, including foods, animal drugs and feeds, human medicines and vaccines, radiation-emitting devices, medical devices, blood and blood products, and cosmetics. With regard to food safety, FDA is responsible under the Federal Food, Drug, and Cosmetic Act for ensuring that all human foods introduced into interstate commerce—except meat, poultry, and processed egg products—and animal feeds are safe, wholesome, and labeled properly. To carry out its responsibilities, FDA has the authority to do such things as conduct examinations and investigations, inspect food facilities, refuse the entry of imported food that appears to be adulterated, and recommend judicial enforcement actions to the Department of Justice. Under the Public Health Service Act, FDA has the authority to take measures, such as issuing regulations, that in its judgment are necessary to prevent the spread of communicable diseases, including foodborne illness.

The Public Health Security and Bioterrorism Preparedness and Response Act of 2002 (Bioterrorism Act) amended both acts. With regard to food safety, the act

- provides FDA the authority to administratively detain a food product where there is credible evidence or information that the product presents a threat of serious adverse health consequences or death to humans or animals;
- directs FDA to issue regulations requiring information regarding food that is being imported or offered for import prior to its arrival at a U.S. port;
- directs FDA to issue regulations requiring domestic and foreign facilities engaged in manufacturing, processing, packing, or holding food for human consumption in the United States to register with FDA; and
- authorizes FDA to issue regulations requiring food firms, except farms and restaurants, to keep records on the immediate previous source and the immediate subsequent recipient of their products.

Within FDA, two centers have primary responsibility for food safety—CFSAN, which is responsible for human food, and the Center for Veterinary Medicine, which is responsible for animal feeds and drugs. In addition, FDA's Office of Regulatory Affairs (ORA) performs food safety and other field work in support of the centers' programs, and the National Center for Toxicological Research conducts food safety research, among other things. CFSAN and ORA perform most of FDA's fresh produce safety activities. CFSAN formulates regulations and guidance, conducts research, communicates information to industry and the public, and sets program priorities, while ORA carries out produce-related field activities, such as conducting inspections, collecting and analyzing samples, and taking enforcement action. Further, both CFSAN and ORA are involved in responding to emergencies involving fresh produce, such as foodborne illness outbreaks, which are coordinated by the Office of Crisis Management within the Office of the Commissioner. In addition, the Associate Commissioner for Foods, also within the Office of the Commissioner, works on food safety issues that may directly or indirectly relate to fresh produce.

Other agencies have responsibilities or programs directly or indirectly related to fresh produce safety:

- CDC within the Department of Health and Human Services conducts surveillance of foodborne illness and provides data and information to other food safety agencies, including FDA.
- Environmental Protection Agency regulates the amount of pesticide that may safely remain on food, including fresh produce.
- National Institutes of Health within the Department of Health and Human Services researches ways to prevent disease, such as foodborne illness.
- USDA conducts food safety research and also supports food safety research, education, and extension programs in the land-grant university system and other partner organizations.

In addition, FDA may enter into formal or informal arrangements with states to do inspections, share resources, or avoid duplication of efforts. Also, states may play a significant role in detecting and responding to outbreaks of foodborne illness.

Produce can become contaminated at any point in the production cycle, and the fact that it is often consumed raw without undergoing a "kill step" that would eliminate pathogens prior to consumption contributes to its potential for causing foodborne illness. At the time of our review, FDA was considering an industry request to allow, among other things, fresh produce to undergo higher doses of irradiation. In August 2008, the agency published a final rule to allow such irradiation for fresh iceberg lettuce and fresh spinach. According to an FDA official, the levels of irradiation that could be tolerated by some types of fresh produce would reduce but not completely eliminate pathogens. Produce grown outdoors is particularly vulnerable to contamination. Some factors that may contribute to contamination include the presence of animals in fields or packing areas, poor water quality, or poor worker sanitation practices. Processing produce into fresh-cut products, such as cut fruits or mixed salads, increases the risk of microbial contamination and growth by breaking down the natural exterior barrier of the produce. If pathogens are present, they can enter the product and then spread into other products being processed at the same time. Produce can also become contaminated after it is harvested and processed, such as during transportation or preparation. For example, produce can become contaminated when it is transported in unclean or improperly refrigerated trucks or when consumers place it on surfaces, such as cutting boards, that have not been thoroughly cleaned after coming into contact with raw meat or poultry. In addition to microbial pathogens, other substances, such as pesticides, may contaminate fresh produce.

According to unpublished FDA data on reported illness associated with FDA-regulated foods, from 1996 through 2006, there were at least 96 outbreaks, 10,253 illnesses, and 14 deaths associated with the consumption of fresh produce.[7] CDC officials told us that available data greatly underestimate the number of foodborne illnesses attributable to fresh produce. Many cases are not reported because the ill person does not seek medical care or the doctor does not take a lab culture. Also, according to CDC officials, in many outbreaks, the pathogen is not identified by state or local public health laboratories because of delayed or incomplete laboratory investigation, inadequate laboratory capacity, or inability to recognize a pathogen as the cause of foodborne illness. Fresh produce and pathogens frequently linked to foodborne illness outbreaks included sprouts (*Salmonella*), leafy greens (*E. coli* O157:H7), tomatoes (*Salmonella*), melons (*Salmonella*), herbs (*Cyclospora*), berries (*Cyclospora*), and green onions (hepatitis A). Once the pathogen is identified, laboratories may send a sample to CDC's PulseNet, a nationwide database that matches pathogen strains. Generally, state and local public health authorities conduct investigations to link the pathogen and the contaminated food. However, CDC may provide assistance. FDA becomes involved when the epidemiology indicates there is an outbreak implicating an FDA-regulated product.

FDA HAS SPENT RELATIVELY FEW RESOURCES ON FRESH PRODUCE SAFETY AND OTHER WORK HAS PREEMPTED FRESH PRODUCE EFFORTS

Although FDA has considered fresh produce a priority area for many years, resource constraints and other work—including counterterrorism efforts and unplanned events such as outbreaks—have caused FDA to delay key fresh produce safety efforts. Because it has no formal program devoted exclusively to fresh produce, FDA allocates resources to fresh produce as part of its overall food safety planning process. Moreover, FDA has not consistently and reliably tracked its spending on fresh produce, thus limiting its ability not only to identify how much it has spent on fresh produce safety but also to plan and manage this spending. Our analysis of FDA estimates and data shows that spending on fresh produce safety was approximately 3 percent of total annual FDA food safety spending in fiscal years 2006 and 2007. Additionally, FDA had few staff solely dedicated to fresh produce.

FDA has Identified Fresh Produce as a Priority for Many Years but Has No Formal Fresh Produce Program

Fresh produce has been a key concern for FDA since at least 1997, when the President announced a national food safety initiative that resulted in several produce-related recommendations, such as developing fast and cost-effective methods for detecting pathogens. That same year, a presidential produce safety initiative called for FDA, among other things, to enhance its oversight of imported produce and develop guidance on good agricultural and manufacturing practices for domestic produce. Additionally, CFSAN has listed produce safety activities as priorities each fiscal year since 1999. Such priorities have included, for example, publishing guidance on fresh-cut produce operations, working with industry to develop good agricultural and manufacturing practices for commodities such as tomatoes and cantaloupes, and conducting initiatives specific to contamination in lettuce and leafy greens. CFSAN also highlighted produce safety as a critical issue in its 2004 produce safety action plan, which identified steps to prevent contamination, minimize public health impacts when contamination occurs, improve communication about fresh produce, and facilitate and support relevant research. Additionally, FDA officials told us that fresh produce safety gained more relevance and prominence within CFSAN as a result of the *E. coli* 0157:117 outbreak in spinach in 2006.

Despite identifying fresh produce safety as a priority area, FDA has no formal program devoted exclusively to fresh produce to which it can allocate resources. Instead, FDA directs resources to fresh produce as part of the process it uses to identify overall food safety priorities, some of which include fresh produce. As part of this process, CFSAN develops and publishes its program priorities each year showing the new and ongoing work it plans to conduct during the next fiscal year. Using these priorities and the President's proposed budget as a starting point, ORA consults with CFSAN to develop its own work plans for carrying out CFSAN's field activities for the upcoming year. ORA also spells out the ideal distribution of field staff, by position and by location, needed to implement CFSAN's planned food safety priorities. However, planned priorities often shift during the year in response to outbreaks of foodborne illness, other emergencies, or resource constraints. Also, FDA's informal hiring freeze from fiscal years 2004 through mid-2007 and lower than expected congressional appropriations have meant some field locations lacked positions needed to implement planned work.

FDA HAS NOT CONSISTENTLY OR RELIABLY TRACKED RESOURCES SPENT ON FRESH PRODUCE SAFETY

In addition to lacking a formal program devoted exclusively to fresh produce, FDA has not consistently or reliably tracked the dollars or staff years it spent on fresh produce safety, thus limiting its ability to plan and manage spending. While the five FDA organizations that conducted food safety work—CFSAN, ORA, the Center for Veterinary Medicine, the National Center for Toxicological Research, and the Office of the Commissioner—were able to provide us with reliable data on their overall food safety spending for fiscal years 2000 through 2007, the three organizations that reported spending resources on fresh produce safety—CFSAN, ORA, and the Office of the Commissioner—could not provide reliable data on fresh produce. Specifically, the systems CFSAN and the Office of the Commissioner use to track their food safety spending do not consistently distinguish fresh produce work from other efforts. Consequently, CFSAN could only provide estimates for the minimum number of dollars and staff years it spent on fresh produce for fiscal years 2006 and 2007 and the Office of the Commissioner could provide no fresh produce spending data or estimates. While ORA was able to track its spending on fresh produce because staff generally enter a code to identify the product and processing method when reporting an activity in ORA's work tracking system, officials acknowledged that not all fresh produce activities are reported as such, and thus ORA also provided estimates of fresh produce spending. Because FDA cannot identify the actual resources it spends on fresh produce, it lacks the information needed to understand whether it is allocating its resources in support of produce safety priorities in the most efficient manner. As a result, FDA's ability to effectively plan and manage its food safety resources is limited.

FRESH PRODUCE HAS BEEN A SMALL PART OF FDA'S FOOD SAFETY EFFORTS

Our analysis of FDA's best available spending estimates shows that fresh produce amounted to at least $18 million in fiscal year 2006 and at least $20 million in 2007, or approximately 3 percent of FDA's total annual food safety spending in each year, as shown in table 1. Similarly, our analysis indicates that FDA spent at least 132 staff years on produce safety in fiscal year 2006 and 130 staff years in 2007, or about 4 percent of its total food safety staff years. Because CFSAN does not require its staff to track work on fresh produce specifically, and instead allows its staff to track fresh produce work under either a general food safety category or a produce-specific category, CFSAN dollar and staff year estimates are minimum amounts. Further, based on the estimates FDA provided, ORA spent the vast majority of FDA's fresh produce resources.

Table 1. FDA Domestic and Imported Food Safety and Fresh Produce Spending, Fiscal Years 2006 and 2007

Dollars in millions

Organization	Fiscal year 2006 Food safety Dollars	Fiscal year 2006 Food safety Staff years	Fiscal year 2006 Fresh produce[a] Dollars	Fiscal year 2006 Fresh produce[a] Staff years	Fiscal year 2007 Food safety Dollars	Fiscal year 2007 Food safety Staff years	Fiscal year 2007 Fresh produce[a] Dollars	Fiscal year 2007 Fresh produce[a] Staff years
Center for Food Safety and Applied Nutrition	$150.3	816	$1.5	10	$157.1	744	$3.6	23
Office of Regulatory Affairs (Field work in support of the Center for Food Safety and Applied Nutrition)	283.5	1,950	16.8	121	296.1	1,793	16.6	106
Center for Veterinary Medicine	54.8	321	0.0	0	58.4	318	0.0	0
Office of Regulatory Affairs (Field work in support of the Center for Veterinary Medicine)	33.0	206	0.0	0	34.6	208	0.0	0
National Center for Toxicological Research[b]	10.3	36	0.0	0	11.5	37	0.0	0
Office of the Commissioner	29.6	184	[c]		31.6	181	[c]	[c]
Total	**$561.6**	**3,513**	**$18.3**	**132**	**$589.1**	**3,281**	**$20.2**	**130**

Source: GAO analysis of FDA data and estimates.

Notes: Totals may not add due to rounding.

[a]According to FDA officials, fresh produce spending is estimated.

[b]Food safety spending includes some dollars spent on collaborative efforts with other federal agencies and external organizations.

[c]The Office of the Commissioner could not provide an estimate of its fresh produce spending because it could not distinguish fresh produce work from other work.

CFSAN reported fresh produce spending in two areas—produce safety and response to foodborne outbreaks—and provided an estimate of related overhead, as shown in table 2. (See app. II for a detailed listing of CFSAN's food safety spending.) The produce safety category includes efforts specific to fresh produce, such as assessing the growing practices and potential contamination pathways in leafy greens and tomatoes. However, food safety work with a fresh produce component, such as an effort to modernize current good manufacturing practice regulations for food, generally falls under the general food safety category. The response to foodborne outbreaks category includes responses to various outbreaks, such as the 2006 outbreak of *E. coli* O157:H7 in spinach. According to CFSAN officials, CFSAN attributed all of its outbreak response spending to fresh produce in fiscal years 2006 and 2007 because fresh produce spending estimates were understated under the produce safety category and outbreak response work—which staff sometimes report more generally as food safety work—primarily involved fresh produce.

Table 2. Minimum Estimates of CFSAN Spending on Fresh Produce, Fiscal Years 2006 and 2007

Activity	Fiscal year			
	2006		2007	
	Dollars	Staff years	Dollars	Staff years
Produce safety	$1,290,696	9.4	$2,330,437	15.9
Response to foodborne outbreaks	31,003	0.2	883,993	6.1
Overhead[a]	176,647	0.7	408,048	1.4
Total	**$1,498,346**	**10.4**	**$3,622,478**	**23.3**

Source: FDA estimated data.

Notes: Totals may not add due to rounding. These estimates represent the minimum resources spent on fresh produce.

[a]Overhead represents CFSAN's portion of shared services, such as human resources, information technology, management services, and telephone expenses, and excludes rent and facilities.

Similarly, ORA reports fresh produce spending across various activities. (See app. II for a detailed listing of ORA's food safety spending.) Based on ORA's estimates for fiscal years 2006 and 2007, it focused most of its produce safety resources on analyzing produce for pesticides and industrial chemicals (49 percent and 56 percent), sampling domestic and imported produce for microbial contamination (26 percent and 12 percent), implementing the general program for domestic food safety (10 percent and 9 percent), and examining imported foods (9 percent and 15 percent). Table 3 shows the dollars and staff years spent on ORA activities involving fresh produce.

Table 3. Estimates of ORA Spending on Fresh Produce, Fiscal Years 2006 and 2007

Dollars in millions				
Activity	Fiscal year			
	2006		2007	
	Dollars	Staff years	Dollars	Staff years
Chemical safety of foods				
Pesticides and industrial chemicals in domestic and imported foods	$8.2	58.9	$9.2	59.0
Radionuclides in foods	<.1	0.1	0.1	0.4
Mycotoxins in domestic and imported foods	<.1	0.1	<.1	0.1
Food and color additives (imports)	<.1	<.1	<.1	0.1
Toxic elements in foods (domestic and import)	<.1	<.1	0.0	0.0
Microbiological safety of foods				
Import and domestic microbiological (produce) assignment	4.4	31.9	2.1	13.2
Audits of food contract inspections	<.1	0.1	<.1	0.1

Table 3. (Continued)

Dollars in millions				
Activity	**Fiscal year**			
	2006		**2007**	
	Dollars	**Staff years**	**Dollars**	**Staff years**
Natural disasters and emergencies	<.1	<.1	0.0	0.0
Nutrient quality and food labeling				
Nutrition Labeling and Education Act, nutrient sample analysis and general food labeling	0.1	1.0	0.1	0.8
Food economics and standards	0.0	0.0	<.1	0.0
Cross-cutting				
Domestic food safety program (general)	1.6	11.7	1.6	9.9
Imported foods (general)	1.5	10.9	2.5	16.2
Emergency response to foodborne outbreaks and illnesses	0.4	3.0	0.6	3.6
Counterterrorism activities	0.3	2.0	0.2	1.5
Consumer complaints	0.1	0.7	0.1	0.5
Foreign inspections and technical assistance	0.1	0.7	<.1	0.2
Short-term assignments initiated by CFSAN, ORA headquarters, or ORA field offices	<.1	0.2	0.1	0.4
Juice hazard analysis and critical control point program	<.1	0.1	0.1	0.3
Miscellaneous other food safety work	0.0	0.0	<.1	<.1
Total	**$16.8**	**121.4**	**$16.6**	**106.2**

Source: GAO analysis of FDA estimated data.

Notes: Totals may not add due to rounding. ORA included overhead in the dollars and staff years spent for each activity.

FDA's Office of the Commissioner also conducts work related to fresh produce, primarily through its Office of Crisis Management, which manages FDA's response to foodborne illness outbreaks and other emergencies. In addition, FDA's Commissioner created a position in 2007 within the Office of the Commissioner responsible for developing a strategic effort to better protect the nation's food supply, which includes fresh produce. However, FDA was unable to provide data or estimates on the resources spent on fresh produce by the Office of the Commissioner.

Additionally, our analysis indicates that FDA had few staff dedicated solely to fresh produce safety. Within CFSAN, two staff have worked almost full-time on fresh produce safety issues for several years, and approximately 25 staff—primarily researchers—have principally worked on fresh produce safety as part of their food safety responsibilities in fiscal year 2007. CFSAN staff have worked on a variety of fresh produce efforts, such as assessing the growing practices for leafy greens and tomatoes, conducting research on contaminants in produce, and working on foodborne illness tracebacks involving fresh produce. While no ORA field staff work solely on produce safety, they do conduct fresh produce-related work as part of other food safety efforts, such as inspecting facilities and testing products under CFSAN's domestic food safety compliance program—a general food safety program focusing on high-risk firms and products not covered by other programs. In addition, ORA has conducted some fieldwork exclusively directed at fresh produce, such as an annual assignment to collect and analyze up to 1,000 samples of domestic and imported fresh produce. While staff within the Office of the Commissioner may work on issues related to fresh produce, such as responding to outbreaks of foodborne illness linked to fresh produce, no staff work solely on fresh produce issues.

CFSAN officials acknowledged that having a critical mass of at least five staff solely working on produce safety, including an expert to lead the group, would be a more effective way of overseeing produce safety. In 2007 and 2008, CFSAN received funding to hire four additional staff to work solely on produce safety. However, CFSAN officials told us that they had not been able to fill all of the positions at the time of our review, in part due to salary limitations.

FDA Has Delayed Key Fresh Produce Activities

Even though fresh produce safety has been a stated priority for many years, other food safety work has preempted produce safety efforts. Specifically, resource constraints and other work—including counterterrorism efforts and unplanned events such as outbreaks of foodborne illness—led FDA to delay key fresh produce safety efforts.

FDA's food safety responsibilities have increased in recent years as a result of the growth in imported food and new regulatory responsibilities, among other things. At the same time, our analysis shows that, after adjusting for inflation, FDA's total food safety spending, which includes fresh produce, remained relatively stable in recent years, despite an initial surge after the events of September 11, 2001. Specifically, FDA's total spending on food safety dropped slightly below its peak of $600.8 million in fiscal year 2003 to $600.3 million in fiscal year 2007 (in constant fiscal year 2008 dollars), as shown in Figure 1. (See app. II for FDA's food safety spending for fiscal years 2000 through 2007.)

At CFSAN, spending on food safety declined slightly from its peak of over $164 million in fiscal year 2003 to approximately $160 million in fiscal year 2007 (in constant fiscal year 2008 dollars). As a result, CFSAN had to absorb cost-of-living increases for its staff, which translated into substantial budget constraints and loss of staff years through early retirements and decisions not to fill vacated positions. CFSAN incurred these losses despite an increase in responsibilities, significantly impairing its ability to fund its priorities, according to the document detailing the center's 2006 program priorities. ORA spending on food safety during this period remained relatively flat (in constant fiscal year 2008 dollars), translating into similar budget constraints and absorption of cost of living increases.

Our analysis also shows that staffing levels for food safety have fallen during the last 4 fiscal years, after initially increasing following September 11, 2001, as shown in Figure 2. (See app. II for detailed information on FDA's food safety staffing for fiscal years 2000 through 2007.) Food safety staffing levels declined 17 percent from their peak of 3,969 staff years in fiscal year 2003 to 3,281 in fiscal year 2007. In fact, the number of staff years FDA spent on food safety had fallen below fiscal year 2002 levels by the end of fiscal year 2007. As a result, FDA lost 70 percent of the food safety staff years gained since fiscal year 2000.

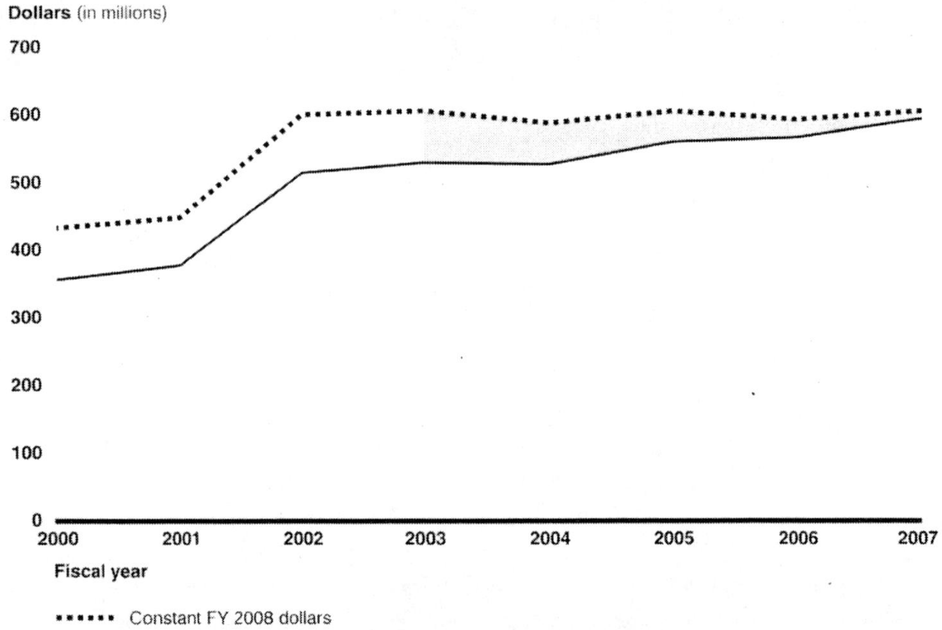

Figure 1. FDA Food Safety Spending in Constant and Nominal Dollars, Fiscal Years 2000 through 2007

Source: GAO analysis of FDA data.

Note: Annual totals include some non-FDA funding from collaborations between the National Center for Toxicological Research and non-FDA entities.

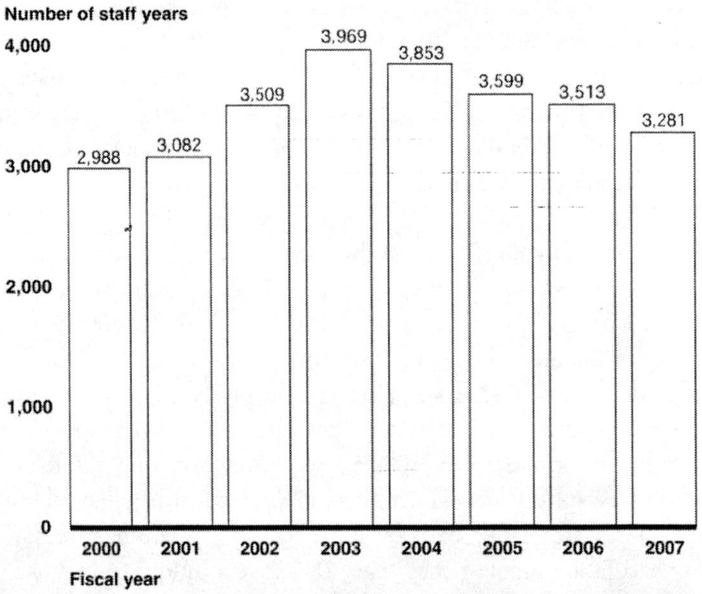

Figure 2. FDA's Total Food Safety Staffing Levels, Fiscal Years 2000 through 2007

Source: FDA data.

Note: Staff years are rounded to the nearest whole number.

Specifically, FDA experienced a net decline of 689 food safety staff years between fiscal years 2003 and 2007.[8] ORA sustained the vast majority of these losses, losing 410 staff years, or more than 20 percent of the staff years it devotes to supporting CFSAN's food safety activities. CFSAN lost 177 food safety staff years during this same period, or almost 25 percent of its headquarters food safety staff years. Figure 3 shows the FDA organizations experiencing a decline in food safety staff years between fiscal years 2003 and 2007, and appendix II, table 5, shows FDA's total food safety spending, by center, during this period.

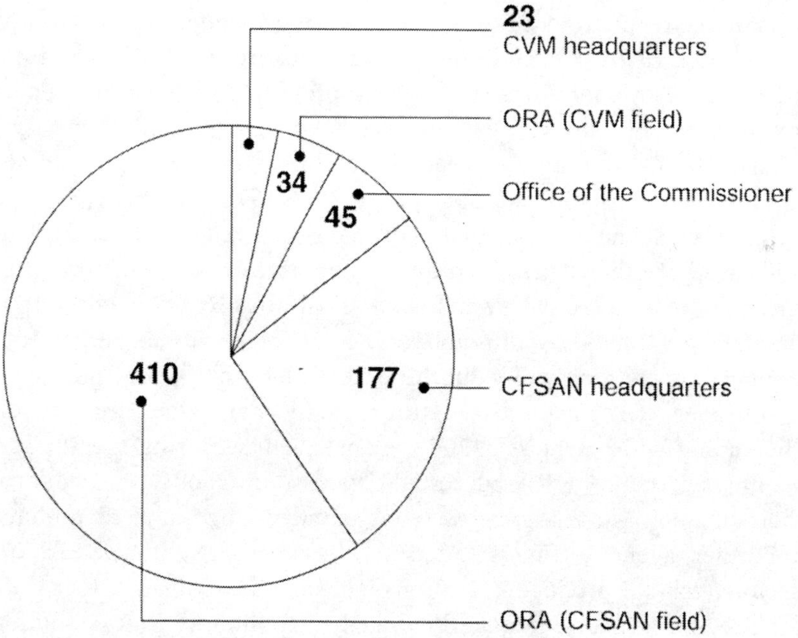

Figure 3. FDA Organizations Experiencing a Decline in Food Safety Staff Years between Fiscal Years 2003 and 2007

Source: GAO analysis of FDA data.

Notes: Staff years are rounded to the nearest whole number. The National Center for Toxicological Research is not included in this Figure because it gained one staff year between fiscal years 2003 and 2007.

In our January 2008 testimony on FDA's *Food Protection Plan*, we noted that FDA's resources have not kept pace with an increasing food safety workload over the past decade and that funding issues are more acute for CFSAN than for other centers because, unlike some other FDA programs, it does not have the authority to collect and retain user fees. Similarly, CFSAN's former director testified in February 2008 that the agency's food-related funding has not kept pace with inflation, with FDA losing 800 scientists, inspectors, and other critical food safety staff since fiscal year 2004. In addition, former Secretaries of the Department of Health and Human Services and other experts have called on the Congress to dramatically increase the agency's budget for protecting the nation's food supply. FDA's Commissioner, in a May 2008 letter to the Congress, assessed the agency as immediately needing an additional \$125 million to achieve its food protection goal for fiscal year 2008. However, as we testified in June 2008, the total costs to fully implement the agency's *Food*

Protection Plan are not yet clear,[9] and we continue to have concerns about the agency's lack of specificity about the resources needed.

FDA officials acknowledged that, in recent years, resource constraints have led FDA to delay work on some key fresh produce safety efforts because FDA lacked enough staff with the needed expertise in areas like regulatory development, as well as enough staff with the technical and subject matter expertise to work on all food safety priorities. For example, according to FDA officials, this staff shortage caused the agency to delay two key efforts relating to fresh produce—modernizing its current good manufacturing practice regulations for food and updating its good agricultural practices guidance. In addition, FDA acknowledged that because CFSAN has lacked the resources to fund its extramural research grant program or perform some critical produce-related research internally, it has not yet been able to conduct the large-scale surveys of fresh produce that are crucial to understanding the incidence of contamination of produce by pathogens such as *E. coli* O157:H7 or *Salmonella*.

FDA officials also acknowledged that they have delayed work on fresh produce safety efforts because of other work, including counterterrorism efforts in response to the Bioterrorism Act and unplanned events such as outbreaks of foodborne illness and recalls of contaminated foods. Additionally, ORA officials repeatedly told us that responding to emergencies, such as foodborne illness outbreaks, takes precedence over other work. However, ORA stopped setting aside staff time and funding for emergencies beginning with fiscal year 2007. Consequently, during the 2007 recall of botulism-contaminated chili, ORA diverted staff throughout the country from conducting their ongoing work to removing potentially contaminated products from store shelves. Further, agency officials stated that the February 2008 release of FDA's final guidance for fresh-cut produce operations—designed to enhance food safety in a segment of the produce industry frequently linked to foodborne illness outbreaks and listed as a CFSAN priority since fiscal year 2000—was delayed at least 6 years because CFSAN had to divert staff with the needed expertise to address counterterrorism efforts and foodborne illness outbreaks.

FDA Has Provided Limited Oversight of Domestic and Imported Fresh Produce

FDA's oversight of domestic and imported fresh produce has generally been limited, with less focus on prevention and intervention—the first two areas of its food safety framework—than on response. This approach has limited the effectiveness of FDA's oversight. First, in terms of prevention, gaps in scientific knowledge and the lack of regulations have limited FDA's efforts to prevent microbial contamination of fresh produce. In terms of intervention, FDA has inspected domestic fresh produce firms infrequently. In addition, although FDA has allocated additional resources to import oversight, it has not been able to inspect a larger share of incoming fresh produce shipments. Finally, FDA gives top priority to responding to emergencies, such as foodborne illness outbreaks, and has had some success in improving its response, but tracing contaminated produce back to its source remains challenging.

Gaps in Science and the Lack of Regulations Have Impeded FDA's Efforts to Take a Prevention-Based Approach

Gaps in scientific knowledge have impeded FDA's efforts to integrate science and risk analysis into its oversight of fresh produce safety. Food safety experts recommend a system that is science-based and uses risk analysis to focus preventive efforts on the foods and processes most likely to cause illness. FDA follows this approach by using available research to inform regulatory decisions and by considering the risk level of different products in making decisions about where to focus resources. However, FDA officials have noted that gaps in science have impeded their ability to make some decisions on how to regulate fresh produce. For example, cattle are known carriers of *E. coli* O157:H7, but scientists do not know exactly how *E. coli* is passed from animals to produce, and thus cannot say how far cattle should be kept from a leafy greens field. Furthermore, FDA does not have sufficient information to develop robust, science-based risk assessments that quantify the relative risks of consuming different types of produce. Lacking such information, FDA largely relies on qualitative information, such as the history of past outbreaks of foodborne illness, to rank the risk levels of fresh produce commodities.

FDA has taken steps to fill some of the gaps in scientific knowledge, but resource constraints have limited the agency's efforts. To fill some of the gaps, FDA conducts laboratory research on fresh produce commodities and their associated pathogens. For example, FDA has a study under way to improve its understanding of how one type of *Salmonella* contaminates tomatoes. However, FDA officials have acknowledged that the scope of their research needs exceeds available resources. Additionally, in response to recurring outbreaks of foodborne illness, FDA implemented a lettuce initiative in 2006 and a tomato initiative in 2007 to study farming practices and environmental conditions that could lead to contamination. These ongoing multiyear initiatives are conducted as part of a risk-based strategy that focuses on growing areas most often linked to past outbreaks—California for lettuce and leafy greens and Virginia and Florida for tomatoes. FDA typically does not inspect farms, so these initiatives also provide an opportunity to assess the extent to which growers are following the agency's recommended practices. FDA intends to use the information obtained through these initiatives to improve guidance and identify additional research or outreach needs. FDA also participates in three research centers in cooperation with academic institutions.[10] However, FDA's Science Board notes that overall output from these centers has been modest because of budget constraints at FDA. Finally, FDA directly funds projects carried out by other institutions, but agency officials noted that resource constraints led the agency to suspend its extramural research grant program for fiscal years 2006 and 2007.

Because FDA has limited resources for food safety research, it relies heavily on other federal agencies, including USDA and the National Institutes of Health, for scientific knowledge. However, a former CFSAN director told us that although FDA has worked to identify research needs and communicate them to other federal agencies, the agency has so far represented a fairly small voice in developing the research agenda, and therefore, gaps in science remain. Additionally, FDA officials acknowledged that it can be difficult to persuade other federal agencies to conduct research that meets FDA's needs. For example, obtaining baseline data on contamination levels of lettuce in different regions and in different seasons would aid FDA's regulatory work, but it is extremely expensive and, according to FDA officials, other agencies are uninterested in funding research to obtain such data. In addition to communicating research needs through meetings and conferences, FDA officials told us that the agency publishes a list of its research needs to communicate its priorities to other federal agencies. However, according to the director of CFSAN's Office of Regulatory Science, the most recent version was last updated 8 years ago, and many of the research priorities identified in that document still have not been addressed. Nevertheless, FDA is currently exploring ways to expand the agency's knowledge on fresh produce safety.

FDA has issued some voluntary guidance for fresh produce but has not issued enforceable regulations to prevent contamination. In 1998, for instance, FDA issued guidance for industry on good agricultural practices for reducing the risk of microbial contamination when growing, packing, and transporting fresh produce. For example, the guidance suggests practices that growers should consider to protect the quality of agricultural water. FDA has also issued guidance specifically for sprouts and for fresh-cut produce, as well as draft guidance for controlling *Listeria monocytogenes* in ready-to-eat foods, including fresh-cut produce. In addition, FDA has provided technical assistance to industry in developing guidance for melons, lettuce and leafy greens, and tomatoes, but the agency has not officially endorsed these documents. While guidance provides useful recommendations, it is voluntary and unenforceable. Finally, FDA has not issued preventive regulations for fresh produce, even though it has done so for other high-risk foods—for seafood in 1995 and for fruit juice in 2001. FDA's current good manufacturing practice regulations for food, which set out basic sanitation rules for food manufacturers, apply to fresh-cut produce, but specifically exempt raw agricultural commodities, including whole fruits and vegetables.

In the absence of preventive regulations issued by FDA, industry groups and others have undertaken other mechanisms to establish additional standards, such as the following:

- Both California leafy greens handlers and Arizona leafy greens shippers have entered into marketing agreements that regulate growing and processing practices for those who sign the agreement.
- USDA issued an advance notice of proposed rulemaking for a national marketing agreement for leafy greens.
- Florida's Department of Agriculture and Consumer Services recently adopted requirements related to safe growing and packing practices for tomatoes.
- The Association of Food and Drug Officials is developing a model fresh produce regulation for use by states.
- Some retail establishments have adopted policies requiring suppliers to meet specific, and often differing, requirements and to demonstrate adherence to these standards through inspections carried out by designated third-party auditors.

While many have praised these efforts as demonstrating a broad base of commitment to enhancing fresh produce safety, some have expressed concern that they lack national uniformity and create a patchwork of requirements that may present problems, both for growers and consumers. To address some of these problems, both industry groups and food safety experts have advocated for federal fresh produce regulations.

FDA's Efforts to Inspect and Sample Domestic and Imported Fresh Produce Have Generally Been Limited

FDA's intervention efforts have been limited. First, FDA infrequently inspected domestic firms handling fresh produce. Additionally, FDA tested fresh produce samples more frequently for pesticides than for microbial contamination. Finally, while FDA has allocated additional resources to import oversight, it has not been able to inspect a larger percentage of imported fresh produce items.

Domestic produce inspections have occurred infrequently. The number of domestic fresh produce inspections conducted by FDA fluctuated from fiscal years 2000 through 2007. Our analysis of FDA data showed that the number of domestic fresh produce inspections rose from 327 in fiscal year 2000 to a peak of 699 in fiscal year 2005, and then declined to 478 inspections in fiscal year 2007. Overall domestic food inspections also fluctuated, rising from about 13,300 in fiscal year 2000 to about 19,800 in fiscal year 2004, and decreasing to about 15,700 in fiscal year 2007, as shown in Figure 4.

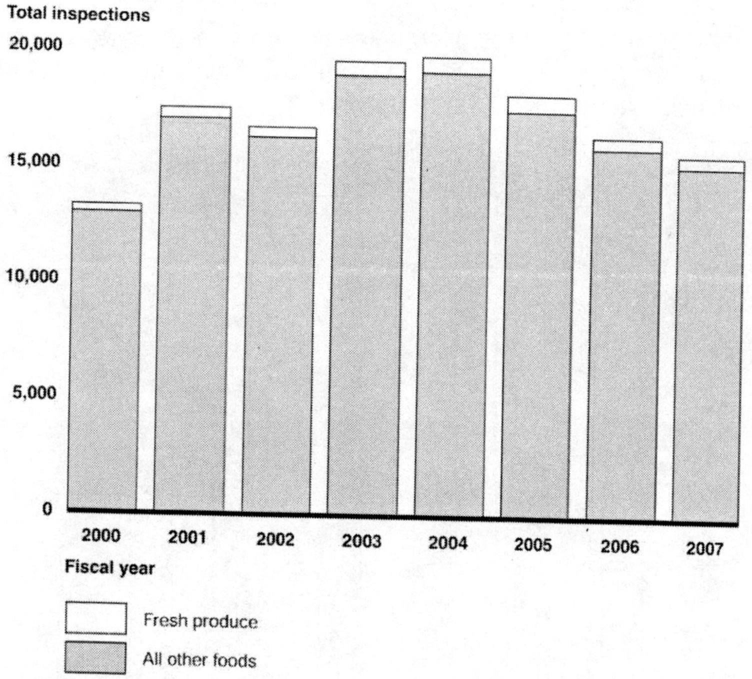

Figure 4. Domestic Food and Fresh Produce Inspections, Fiscal Years 2000 through 2007

Source: FDA data.

We also found that firms were inspected infrequently. Because FDA's database of registered food firms does not capture data that would identify all U.S. firms that handle fresh produce, we could not determine the percentage of fresh produce firms inspected.

However, according to our analysis, 2,002 firms had at least one produce-related inspection between fiscal years 2000 and 2007. On average, each firm was inspected twice during that period. About half of the firms had only one fresh produce-related inspection, and only 6 firms (0.3 percent) had a fresh produce-related inspection every year.

Additionally, FDA has increasingly coordinated with states to better leverage inspection resources. At the end of fiscal year 2007, FDA had contracts with 40 states to conduct food inspections on its behalf and had less formal partnership agreements with some states to coordinate food safety activities, including inspections. We found that states conducted 200 of the 600 fresh produce inspections (33 percent) in fiscal year 2003—including one conducted jointly with FDA officials—compared with 269 of the 478 (56 percent) in fiscal year 2007, as shown in Figure 5. FDA officials attributed this increase in the share of inspections conducted by states to an overall effort to better leverage resources in response to FDA staff losses and resource constraints. FDA officials also noted that the agency has tried to maintain consistent funding for state-conducted inspections. As a result, when changes in FDA funding occurred, the number of FDA-conducted inspections fluctuated more than state-conducted inspections. State officials noted that increasing the share of domestic inspections performed by the states may allow FDA to concentrate on areas where states do not have jurisdiction, such as import oversight. However, state officials also noted that they have experienced some challenges in partnering with FDA. For example, according to one state official, FDA sometimes restricts state access to the agency's inspection results because of confidentiality concerns. According to the official, such data would provide states with useful information on a firm's history and help the states' inspection efforts.

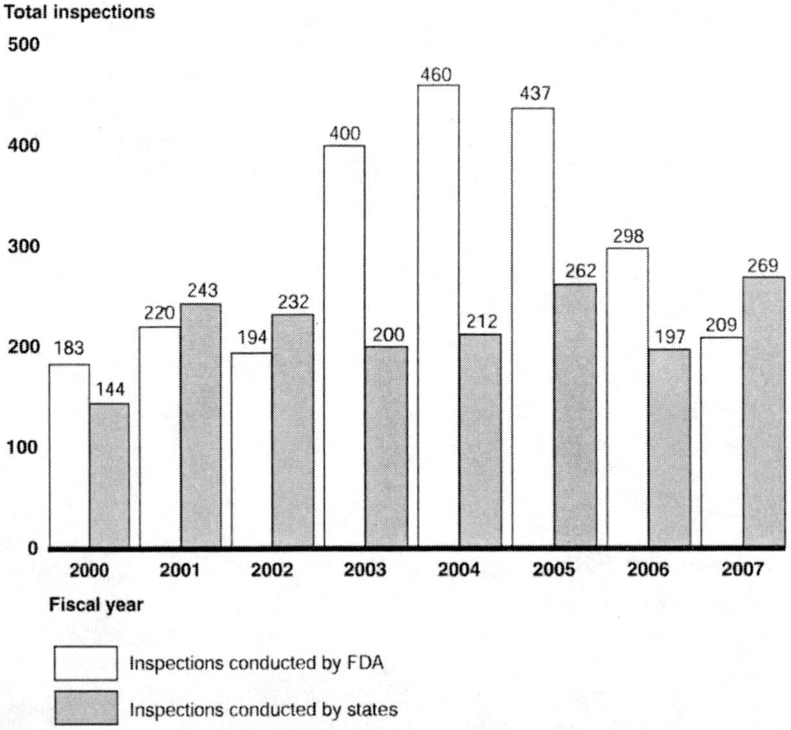

Figure 5. Fresh Produce Inspections Conducted by FDA and States, Fiscal Years 2000 through 2007

Source: GAO analysis of FDA data.
Note: State-conducted inspections include seven inspections conducted jointly with FDA.

FDA and the states found problems in approximately 41 percent of the fresh produce inspections conducted between fiscal years 2000 and 2007, as shown in Figure 6. When inspections find objectionable conditions, it is FDA's practice to give firms an opportunity to take voluntary corrective action before initiating an enforcement action, unless there is significant impact on public health. FDA told us that when an inspection uncovers objectionable conditions, the firm should be reinspected within 2 to 3 years. While FDA noted that the agency is often unable to meet this goal, it also noted that serious problems that warrant regulatory action are likely to trigger a reinspection. More specifically, we were told the agency prioritizes follow-up inspections based on the severity of problems found and a firm's past history of compliance.

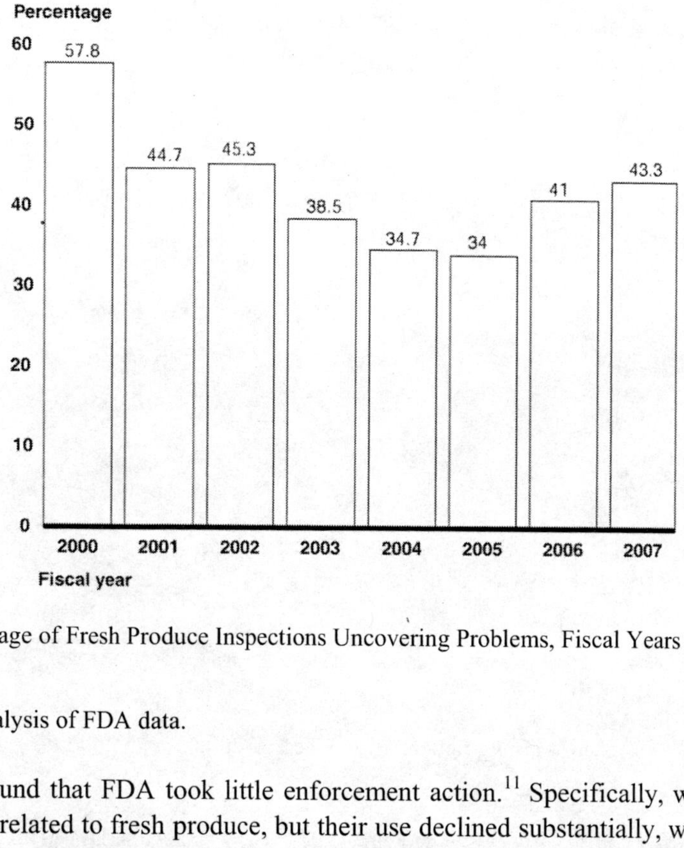

Figure 6. Percentage of Fresh Produce Inspections Uncovering Problems, Fiscal Years 2000 through 2007

Source: GAO analysis of FDA data.

We also found that FDA took little enforcement action.[11] Specifically, we identified 96 warning letters related to fresh produce, but their use declined substantially, with 66 issued in fiscal year 2000 and none issued in fiscal years 2005 through 2007, as shown in Figure 7. During the same period, according to FDA, the agency seized no fresh produce, sought no injunctions, and prosecuted no firms for fresh produce-related violations. We could not determine the number of less formal enforcement actions taken at the FDA district level, such as sending untitled letters—informal communications used to notify firms that corrective actions are needed—or holding meetings to inform firms of objectionable conditions, because FDA had not centrally compiled such data.

Most of FDA's sample analysis has focused on pesticides. In fiscal year 2007, 82 percent of all fresh produce samples were tested for pesticides, while 17 percent were tested for microbial contamination. However, the relative rates of testing differed for domestic and imported produce. FDA officials told us that in the late 1990s, outbreaks of foodborne illness led the agency to focus its domestic food safety program more heavily on microbial contamination, and as a result, the focus of its domestic sample analysis also began to shift. Specifically, in fiscal year 2000, we found that FDA tested 14 percent of domestic produce samples for microbial contamination, compared with 37 percent in fiscal year 2007. According to FDA officials, pesticides remain a larger concern for imported produce, so a significant portion of import testing continues to focus on pesticides. For example, in fiscal year 2007, 92 percent of imported produce samples were tested for pesticides. FDA officials told us that two factors have slowed the agency's shift toward microbial testing. First, pesticides are easier to detect because they are fairly evenly distributed across products, whereas microbial contamination may be sporadic. Second, pesticide testing can be more productive because available methods allow FDA to test one sample for multiple pesticide residues, whereas similar methods are considerably more difficult in microbial testing.

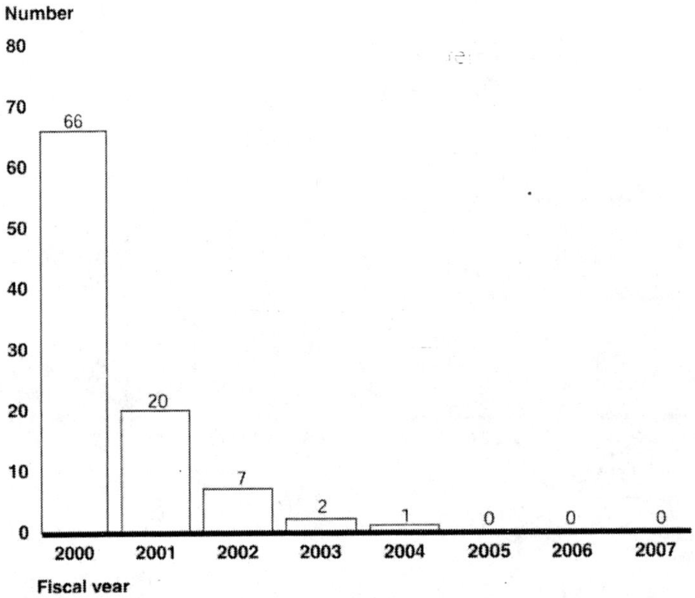

Figure 7. Fresh Produce-Related Warning Letters, Fiscal Years 2000 through 2007

FDA has provided limited oversight of imported fresh produce. FDA primarily relies on an electronic screening process to review imported produce at the border—typically only inspecting foreign produce firms for cause, such as a potential link to an outbreak of foodborne illness. The basic import process consists of two stages—prior notice and food safety evaluation. In the first stage, FDA must receive prior notice before a food shipment arrives in the United States. Prior notice information is screened electronically by FDA's import database, the Operational and Administrative System for Import Support (OASIS), for potential risks associated with intentional contamination. Once the prior notice review has been completed, the food safety evaluation is conducted. For this evaluation, OASIS screens each entry line—or portion—of the shipment for risk factors associated with unintentional contamination to determine whether the shipment may proceed automatically or whether it requires further review.[12] In fiscal year 2007, about one-quarter of all fresh produce entry lines received a system "may proceed" designation after this step. If an entry line does not receive this designation, an FDA reviewer conducts an on-screen evaluation. OASIS alerts the reviewer to factors that may prompt further action, such as the existence of an import alert or a product meeting the criteria for sampling, and the reviewer then decides whether or not to allow the entry line to proceed. According to an FDA official, in 2007 this on-screen review took an average of about 45 seconds, and the vast majority of entry lines are allowed to continue into domestic commerce. In fiscal year 2007, nearly 99 percent of all fresh produce entry lines were cleared following electronic screening. Entry lines that are held by FDA may be physically examined, sampled, or detained without physical examination.

Food imports in general and fresh produce imports in particular have increased in recent years, and despite allocating additional resources, FDA has not been able to inspect a larger portion of fresh produce imports. Specifically, imported food for human consumption increased 84 percent from fiscal years 2002 through 2007, from about 5.1 million lines to about 9.4 million lines. Fresh produce imports increased about 60 percent over the same period, from about 940,000 lines in fiscal year 2002 to over 1.5 million lines in fiscal year 2007. During this period, the number of fresh produce lines that FDA examined more than quadrupled. Specifically, following the passage of the Bioterrorism Act in 2002, the agency received additional resources to oversee both domestic and imported food, which allowed it to increase the number of imported fresh produce lines examined in fiscal year 2003. Additionally, recognizing the substantial growth in imported products, the agency decided around fiscal year 2004 to shift additional resources to import oversight, according to an FDA official. Doing so enabled FDA to increase further the number and percentage of fresh produce lines it examined, but it has not been able to sustain its inspection rate at the fiscal year 2005 level. Although FDA increased fresh produce examinations from about 9,000 lines in fiscal year 2004 to about 11,000 in fiscal year 2007, the percentage of all fresh produce lines inspected remained at about 0.73 percent, as shown in table 4. Overall, FDA examined 0.77 percent and sampled 0.22 percent of the 7.6 million fresh produce lines imported from fiscal years 2002 through 2007.

Table 4. Actions Taken on Fresh Produce Entry Lines, Fiscal Years 2002 through 2007

Percentage of total lines in parenthesis						
Description	**Fiscal year**					
	2002	**2003**	**2004**	**2005**	**2006**	**2007**
Total import lines	941,365	1,074,076	1,249,645	1,359,978	1,432,316	1,503,884
Lines examined	2,497	8,439	9,106	16,583	10,674	11,014
	(0.27)	(0.79)	(0.73)	(1.22)	(0.75)	(0.73)
Lines sampled	2,879	2,681	2,791	3,204	2,321	2,967
	(0.31)	(0.25)	(0.22)	(0.24)	(0.16)	(0.20)
Lines detained without physical exam	1,822	1,770	2,161	1,796	2,038	1,807
	(0.19)	(0.16)	(0.17)	(0.13)	(0.14)	(0.12)

Source: GAO analysis of FDA data.

Note: Percentages may be slightly overstated because some lines may have been examined more than once or detained for more than one reason and, therefore, double-counted in our annual totals.

According to FDA officials, import alerts are the agency's primary mechanism for keeping products with a history of violations out of the country, and they use them regularly. Through the use of import alerts, the agency may detain potentially adulterated products at the border without a physical exam. Additionally, import alerts place the burden on the importing firm to demonstrate that the product is safe. However, we found that import alerts covered a very small percentage of fresh produce imports from fiscal years 2002 through 2007. During that period, FDA detained 0.15 percent of all imported fresh produce lines on the basis of import alerts. Officials attributed the small proportion of imports covered by import alerts to the fact that FDA only samples a small portion of imports and only issues an import alert after it finds a problem.

FDA Has Taken Steps to Improve Response to Outbreaks and Other Emergencies but Continues to Face Challenges

FDA has had some success in improving its response to produce-related emergencies but continues to face challenges. In particular, FDA has had success in improving coordination, partnering with California on outbreak investigations, and piloting a program involving recalled products. While FDA has taken steps to improve its ability to trace outbreaks to their source, it remains extremely difficult to do so for fresh produce. Additionally, FDA lacks authorities that could be useful in responding to food-related emergencies, including mandatory recall authority for foods other than infant formula, and access to certain records during emergencies.

FDA has taken some steps that have helped to improve the agency's response to produce-related emergencies. For example, in 2002, FDA established an Office of Crisis Management within the Office of the Commissioner. Creating this office has enhanced the agency's ability to coordinate its emergency response activities internally and with others. Also, realizing that they often performed parallel investigations for the same outbreak, FDA's Pacific Region and the California Department of Public Health decided to leverage resources by developing a joint emergency response group, called the California Food Emergency Response Team (CalFERT). CalFERT includes highly skilled and experienced investigators and scientists from FDA and California who regularly train together, allowing them to develop a working relationship before an emergency occurs. Both FDA and state officials said that CalFERT enabled them to respond quickly to the 2006 *E. coli* O157:H7 outbreak linked to spinach. Finally, in order to improve consumer awareness of recalled products, FDA started a pilot program in 2007 in which they include a photograph of the label of a recalled food product on their Web site along with the announcement of the recall. FDA is currently evaluating this program, but initial results indicate that it has been effective in helping consumers identify recalled foods.

FDA has also taken steps to improve its ability to trace outbreaks to the source of contamination, but it remains extremely difficult to do so for fresh produce. Traceback investigations are an important part of emergency response because they help identify which products should be recalled and how contamination occurred. The traceback process generally starts at the retail locations implicated in cases of illness and follows the implicated product back through the supply chain to identify a common source.

Such investigations can be particularly difficult when they involve fresh produce because produce is highly perishable and may no longer be available for testing when an outbreak is identified. Even when products are available, they often are unlabeled or mixed in packages containing products from multiple sources, making it difficult to identify a specific source of contamination. For example, tomatoes can be very difficult to trace because packing houses often combine shipments from multiple domestic and foreign growers in order to create boxes of similar quality or size. To address some of these challenges, FDA has developed traceback procedures specifically for fresh produce investigations that provide techniques for determining and documenting the distribution and production chain and the source of the product implicated in an outbreak. FDA has also developed farm investigation procedures for cases where contamination may have occurred in the field. Additionally, FDA offers a course to help familiarize FDA and state inspectors with outbreak investigation procedures on farms. Despite these efforts, traceback remains very difficult for fresh produce, and it remains uncommon for investigations to trace the product back to the field that was likely the source of contamination. CalFERT was able to do so following the 2006 outbreak of *E. coli* O157:H7 in spinach because contaminated product was found in bags with lot numbers.

However, FDA lacks some authorities that could improve its response efforts, including mandatory recall authority for foods other than infant formula and the authority to access some firm records during emergencies. FDA provides assistance to industry during voluntary recalls but does not have the authority to require a firm to issue a recall for any food other than infant formula. When industry issues a recall, FDA assists firms in notifying distributors and consumers, classifies the recall based on the potential threat to human health, and monitors the firm's recall process. Fresh produce recalls are voluntary, and FDA officials acknowledge that firms can ignore an FDA request to initiate a recall. Additionally, while FDA has the authority to access certain records, including processing and shipping records, for a product it reasonably believes is adulterated or presents a threat of serious adverse health consequences or death, it does not necessarily have the authority to access records for related products handled in the same facility. FDA officials acknowledged that they sometimes obtain records from states, since some have broader authority to access firm records. However, having to ask states to obtain records could slow an outbreak investigation.

Proposed Actions Could Significantly Enhance Fresh Produce Oversight, But More Information Is Needed to Assess the Likelihood of Success

Through the *Food Protection Plan*, FDA has proposed actions and identified additional authorities that could significantly enhance its oversight of fresh produce within the plan's framework of prevention, intervention, and response. However, FDA has not provided sufficient information on strategies and resources for implementing the plan, which makes it difficult to assess the likelihood of success.

FDA's Plans to Help Fill Gaps in Scientific Knowledge and Issue New Guidance and Regulations Could Enhance Prevention Efforts

To help prevent contamination of fresh produce, FDA plans to help fill gaps in scientific knowledge and issue new guidance and regulations, and has cited in its *Food Protection Plan* a need for explicit authority to issue regulations that are intended to prevent contamination. Specifically, FDA intends to:

- *Help expand scientific information on fresh produce safety*. FDA plans to expand knowledge on fresh produce safety. For example, FDA is working with researchers from several universities to carry out a USDA-funded project looking into options for reducing the risk of *E. coli* O157:H7 in leafy greens from production to packaging. Also, in June 2008, FDA announced the establishment of the Western Center for Food Safety at the University of California, Davis, which will create a research, education, and outreach program focused on the interface between agriculture and food safety. The center plans to focus its initial research on produce safety, such as safe agricultural practices for domestic and imported commodities. FDA also plans to strengthen its current qualitative risk ranking of food commodities and pathogens, starting with fresh produce items. In addition, FDA announced in June 2008 that it will fund approximately $1 million in extramural research grants that address fresh produce safety topics, such as how consumer handling of fresh-cut produce may compromise microbiological safety and problems that occur during transportation. To help shape outside research and update its 8-year-old list of research needs, FDA officials told us that CFSAN is developing a plan that outlines priority research needs, including produce safety research. According to FDA officials, CFSAN has just begun to develop this plan, and it may take several months before it is available. Finally, FDA officials told us that they would like to have voluntary access to data from producers for research purposes. For example, some fresh produce firms have testing records that show when they have found *E. coli* O157:H7 or Salmonella in product samples. FDA officials noted that they are interested in conglomerate data stripped of identifiers, but that in order to use industry data, they would need to find ways to effectively address industry members' reluctance to share such information.

- *Update existing fresh produce-related guidance and regulations*. FDA officials told us they plan to update the agency's 1998 good agricultural practices guidance to incorporate new knowledge about safe growing practices for fresh produce. FDA also has plans to update its current good manufacturing practice regulations for food, which were last updated in 1986 and which guide, among other things, domestic inspections of fresh-cut produce facilities. While both efforts could enhance FDA's oversight and assist industry in producing safe produce items, at the time of our review, FDA officials said they had not yet started to update the agency's good agricultural practices guidance because of limited resources. Furthermore, progress on updating the current good manufacturing practice regulations for food has been slow since the effort began in 2002, and officials could not provide an estimate of their completion date.

- *Seek authority to issue preventive safety regulations for foods such as fresh produce.* FDA, in the *Food Protection Plan*, identified a need for explicit authority from the Congress to issue regulations to require preventive measures by firms producing foods that have been associated with repeated instances of serious health problems or death. FDA already has preventive regulations for seafood and juice, which require firms to analyze safety hazards and implement plans to address those hazards. According to FDA, such authority would strengthen the agency's ability to implement risk-based processes to reduce illnesses from high-risk foods. We have previously recommended the use of such preventive safety regulations for chemicals,[13] and FDA officials told us that issuing preventive regulations may be one of the most important things they can do to enhance their oversight of fresh produce. While some consumer groups, food safety experts, and producers agree that uniform standards could enhance safety for targeted foods, others have noted that preventive regulations can take a long time to develop and it may be difficult to incorporate new scientific knowledge into regulations. FDA officials expressed concern about the public health benefit of having regulations without the resources to enforce them. FDA could not provide us with information on projected costs associated with such regulations, but FDA's Science Board noted that modernizing safety standards for fresh produce and other raw foods, and implementing accompanying inspection programs could total $210 million.

FDA Has Planned Several Actions to Enhance the Effectiveness of Its Intervention Efforts

To enhance the effectiveness of its intervention efforts, FDA plans to use more systematic, risk-based criteria to more formally target domestic inspections and introduce a new import screening tool and has identified a need for authority to accredit third parties. All these efforts have the potential to increase the effectiveness of fresh produce oversight and better leverage resources. Specifically FDA intends to:

- *Use more information to target domestic inspections.* FDA plans to enhance its risk-based criteria for determining which facilities to inspect by including more information, such as a firm's previous inspection results, recalls, and association with outbreaks and adverse events. FDA officials told us that they have already begun to use some new risk-based information to plan fiscal year 2009 inspections and that, as new information from FDA's ongoing risk-ranking efforts becomes available, they will continue to incorporate that as well.

- *Use a new screening tool for imports.* FDA officials are testing a software tool called Predictive Risk-based Evaluation for Dynamic Import Compliance Targeting (PREDICT) that uses information sources and automated data mining techniques not available to OASIS to detect possible problems in import shipments. For example, PREDICT can use up-to-date data from FDA laboratory tests and other information, such as weather events that could affect water quality, to identify import entries for further review. PREDICT was pilot-tested on seafood imports in mid-2007 in FDA's Los Angeles district office, and the pilot test received a positive evaluation. Former FDA officials have praised PREDICT for its potential to enhance overall food safety oversight, and FDA officials noted that the tool may be particularly useful for fresh produce imports because PREDICT can make use of the agency's testing data for imported fresh produce, which is not currently used in FDA's electronic screening. FDA officials plan to increase the number of high-risk food products and test PREDICT at a border crossing by March 2009.

- *Seek authority to accredit third parties.* Obtaining the authority to accredit third parties could help FDA leverage its inspection resources, and FDA officials told us they believed third-party inspections would be especially helpful for imports, as FDA inspects few foreign food firms. However, FDA does not envision accrediting third parties for fresh produce inspections until the agency has established or assessed the adequacy of clear preventive standards. FDA has already taken this approach by accrediting third parties to inspect manufacturers of medical devices, as authorized by the Congress, and we recently reported that few inspections have been conducted under FDA's accredited programs.[14] However, FDA officials told us that an inspection program for foods could be more attractive because importers and consumers might be more willing to buy products certified by FDA-approved third parties, and third-party certification might expedite FDA review and the entry of perishable commodities into the United States. While third-party inspectors could provide several benefits, FDA officials told us that before consistent, reliable third-party inspections would be feasible for fresh produce, the agency would first need to establish or assess the adequacy of clear preventive standards. For this reason, they told us that FDA would likely use a third-party inspection system for items such as seafood and juice before expanding a program to fresh produce. While USDA has expressed concern that third-party inspections paid for by the companies whose facilities were being inspected may not provide effective enforcement for noncompliant firms, FDA officials noted that the use of accredited third-party inspections would be voluntary and that FDA would not waive any of its authorities.

FDA Has Efforts Under Way That Could Improve Its Response to Emergencies

FDA has efforts under way to enhance its response to food-related emergencies, including enhancing traceback investigations, organizing more federal-state emergency response teams, and improving risk communications. The agency has also identified the need for mandatory recall authority. These efforts could increase the overall effectiveness of FDA's response to outbreaks linked to fresh produce. Specifically, FDA plans to:

- *Enhance traceback efforts.* FDA officials told us they have several efforts under way to enhance traceback investigations, including establishing a working group and purchasing new laboratory equipment. Also, the *Food Protection Plan* identifies a need for enhanced access to firm records. First, the agency has established a working group that has been meeting with industry groups and consumers to gather information and then report on key elements of effective traceback systems. Using this information, FDA plans to develop guidance for industry on traceback systems. Also, FDA has purchased equipment that will enable the agency to more rapidly screen for and accurately identify variants of *Salmonella* in fresh produce and other foods. Being able to identify *Salmonella* variants will enhance FDA's ability to investigate outbreaks. Last, FDA stated in its *Food Protection Plan* that it needs additional authorities to provide improved access to companies' records in a food-related emergency to help the agency quickly identify sources of contamination and take action.

- *Organize more federal-state emergency response teams.* FDA hopes to develop additional federal-state emergency response teams. At the time of our review, FDA was soliciting interest from states and hoped to start developing teams in six states in fiscal year 2009. According to FDA officials, training is an important part of developing these teams, and FDA is working with the Western Institute for Food Safety and Security to create an effective training program for participating states. Officials noted, however, that one of the reasons for the success of CalFERT during the 2006 outbreak was the high level of expertise of CalFERT officials and that similar results may not always be replicable in other states.

- *Improve risk communications.* In February 2008, FDA first convened an expert committee called the Risk Communication Advisory Committee to provide information on ways for FDA to enhance communications during emergencies, such as foodborne illness outbreaks linked to fresh produce. FDA also has plans to conduct studies on consumer communications. Using this information, FDA plans to update its risk communications plan. While these actions could improve FDA's methods for sharing information, the advisory committee has only met twice, and other actions have not yet been completed, so it is too early to tell what effect they will ultimately have.

- *Seek authority to order a recall.* In the *Food Protection Plan*, FDA identified the need for authority to order a recall when FDA has reason to believe that food is adulterated and presents a threat of serious adverse health consequences, which would be imposed only if a company refuses or unduly delays a voluntary recall. Currently, FDA does not have the authority to compel companies to recall contaminated food other than infant formula. FDA does have the authority, through the courts, to seize, condemn, and destroy adulterated or misbranded food and to disseminate information about food items that are believed to present a danger to public health. We have previously noted that limitations in FDA's recall authority heighten the risk that unsafe food will remain in the food supply.[15] In the case of fresh produce, FDA officials told us they were aware of no case where a fresh produce firm refused FDA's recall request, and a former FDA official told us that mandatory recall authority may be less useful for fresh produce items that are highly perishable and may no longer be in commerce by the time FDA becomes involved in an outbreak response.

Limited Information on Strategies and Resources Makes It Difficult to Assess the Likelihood of FDA's Success

FDA issued its first progress report in July 2008 on its 2007 *Food Protection Plan*, and FDA officials told us that the plan has already had a positive effect on food safety by helping to increase the visibility of food protection during the agency's internal budget process. It has also aided FDA in prioritizing its efforts and has led to greater coordination within FDA to address issues that cut across different FDA centers. We have noted that public reporting is the means through which the federal government communicates the results of its work to the Congress and the American people. Such reporting is in the public interest and promotes transparency in government operations. The *Food Protection Plan* contains overarching goals to prevent foodborne contamination, intervene at critical points in the food supply chain, and respond rapidly to minimize harm. However, FDA has provided limited information on strategies and resources, making it difficult to assess the likelihood of achieving these goals. While FDA has released information on spending priorities on its short and medium term efforts for fiscal years 2008 and 2009, it has still not publicly provided information on the full costs of implementing the *Food Protection Plan* or committed to timelines for implementing produce-focused elements of the plan. Information on strategies and resources is increasingly critical, given that important elements of the plan could be highly resource-intensive. This lack of information, coupled with FDA's resource constraints and slow progress on some food protection efforts, such as updating good agricultural practices guidance or issuing guidance for fresh-cut produce operations, decreases public accountability and raises concerns about whether FDA will have the capacity to fully implement the plan.

Additionally, the *Food Protection Plan* recognizes that FDA needs to partner with the Congress to make the changes necessary to transform the safety of the nation's food supply, including legislative changes to strengthen FDA's ability to protect Americans from foodborne illness. However, as we testified in June 2008, FDA's congressional outreach strategy is general. For example, when we asked FDA officials if they had a congressional outreach strategy, they told us that they had met with various congressional committees to discuss the *Food Protection Plan*. When asked if they had provided draft language to congressional committees on the various authorities, FDA officials explained that they only provided technical assistance, such as commenting on draft bills, to congressional staff when asked.

CONCLUSIONS

Fresh produce is essential to a healthy diet and to the health of the industry that produces it. However, fresh produce poses particular safety challenges because it is often consumed raw without any type of treatment that would reduce or eliminate pathogens prior to consumption. FDA plays a critical role in ensuring the safety of fresh produce, yet it has struggled to fulfill that role because of resource constraints, gaps in science, and lack of legal authorities. Specifically, FDA last set its research priorities 8 years ago and has not systematically worked with others to supplement its research agenda, including research relating to fresh produce. FDA's 1998 good agricultural practices guidance has not been updated, and its current good manufacturing practice regulations for food, which includes fresh-cut produce operations, was last revised in 1986. Also, through the *Food Protection Plan*, FDA has proposed actions and identified a need for additional authorities related to preventive controls for high-risk foods and access to certain records that could significantly enhance the agency's oversight of fresh produce. However, the timelines and resources needed to fully implement the plan are unclear. To increase congressional and public confidence and fulfill its mission of protecting public health, it is imperative that FDA follow through on its planned actions to enhance fresh produce oversight, seek needed authorities from the Congress, and foster transparency and accountability by providing specific information to the Congress and to the public on strategies and resources for implementing its *Food Protection Plan*.

Recommendations for Executive Action

We are making seven recommendations to the Commissioner of FDA.

To enhance FDA's oversight of fresh produce safety, we recommend that the Commissioner of FDA see that the agency takes the following actions:

- develop a plan for identifying research priorities and facilitating research related to fresh produce;
- identify approaches for obtaining testing and other information from industry members to inform its research agenda;
- update its good agricultural practices guidance for fresh produce to incorporate new knowledge about safe growing practices; and
- update its current good manufacturing practice regulations for food to incorporate new knowledge about the food industry and safe manufacturing, processing, and holding practices.

To enhance FDA's authority to oversee fresh produce, we recommend that the Commissioner of FDA seek authority from the Congress to

- make explicit FDA's authority to adopt preventive controls for high-risk foods, and
- provide FDA enhanced access to firm records during food-related emergencies.

To foster transparency and accountability, we recommend that the Commissioner of FDA provide specific information to the Congress and to the public on the strategies and resources for implementing the *Food Protection Plan*.

Agency Comments and Our Evaluation

We provided a draft copy of this chapter to the Department of Health and Human Services for review and comment. We received a written response from the Assistant Secretary for Legislation that included comments from FDA. FDA generally agreed with the report's accuracy and conclusions and appreciated our recognition of the *Food Protection*

Plan as a sound framework for advancing food safety and food defense and our use of the plan as an organizing feature for this chapter. FDA generally agreed with five of the report's recommendations and disagreed with two others. FDA's comments and our detailed responses are presented in appendix III of this chapter.

FDA agreed with our recommendation to develop a plan for identifying research priorities and facilitating research related to fresh produce. FDA said that both CFSAN and the agency, as part of the *Food Protection Plan*, were developing strategic plans for research, including fresh produce-related research. FDA said that CFSAN's plan will identify regulatory research priorities that can be addressed through intramural and extramural research, as well as future research needs that cannot be addressed due to resource limitations.

FDA also agreed with our recommendation to identify approaches for obtaining testing and other information from industry members to supplement its research agenda. FDA noted, however, that the data and information from industry would further inform, rather than supplement, the agency's research agenda and would also be used in agency risk assessments associated with fresh produce. We revised the recommendation accordingly.

FDA agreed with our recommendation to update its good agricultural practices guidance for fresh produce to incorporate knowledge about safe growing practices. FDA said the agency recognized, when it issued the guidance in 1998, that it would need to be updated in light of new information and technological advances. FDA added that the agency has completed or is doing many of the actions necessary to update the guidance. FDA stated that it will update the guidance once it evaluates the data and information collected through these efforts.

FDA also agreed with our recommendation to update its current good manufacturing practice regulations for food to incorporate new knowledge about the food industry and safe manufacturing, processing, and holding practices. FDA said that a working group has examined the regulations and identified those areas where risk-based preventive controls would have the greatest impact on ensuring food safety.

While FDA agreed with the importance of having explicit authority to adopt preventive controls for high-risk foods and having enhanced access to firm records during food-related emergencies, the agency believes that it has already sought such authorities by outlining legislative needs in the *Food Protection Plan* and testifying on the plan before congressional committees. We do not view these actions as seeking authority. Rather, as FDA recognized, there is a need for the agency to partner with the Congress to make the necessary changes to transform the safety of the nation's food supply. FDA should move beyond outlining needs and continue to take steps to obtain these legislative authorities such as by suggesting language that provides FDA the necessary statutory tools to help the agency conduct its oversight responsibilities. FDA proposed that these recommendations be redirected to the Congress for action and refers to a prior recommendation we made that the Congress enact comprehensive, uniform, and risk-based food safety legislation. We reiterated this recommendation to the Congress most recently in our High-Risk Series[16] as one action that can help address the fragmented federal oversight of food safety and integrate the myriad food safety programs. Our recommendations to FDA are intended to enhance FDA's authority to oversee fresh produce for which FDA has primary responsibility. We stand by these recommendations.

Finally, FDA generally agreed with our recommendation to provide specific information to the Congress and to the public on the strategies and resources for implementing the *Food*

Protection Plan. FDA agreed with the need for transparency and accountability and noted that the agency has provided information to the Congress on the implementation of the *Food Protection Plan*. In addition, we believe that FDA should publicly release information on strategies and resources, including information on the full costs of implementing the *Food Protection Plan* and timelines for produce-focused elements of the plan. As stated in our report, such information would help the public assess the likelihood of achieving the goals stated in the *Food Protection Plan*.

We are sending copies of this chapter to interested congressional committees, the Secretary of Health and Human Services, and the Commissioner of the Food and Drug Administration.

Lisa Shames
Director, Natural Resources
and Environment

APPENDIX I: OBJECTIVES, SCOPE, AND METHODOLOGY

This chapter examines (1) the dollars and staff years the Food and Drug Administration (FDA) has spent on fresh produce safety and how FDA has allocated those resources, (2) the effectiveness of FDA's actions to oversee domestic and imported fresh produce safety, and (3) the actions FDA plans to take to enhance fresh produce oversight and the extent to which FDA's planned actions address identified challenges. For this chapter, fresh produce means fruits and vegetables in their unpeeled, natural form, as well as fruits and vegetables that have been minimally processed (e.g., peeled, sliced, or chopped), with or without washing, before being packaged for use by the consumer or a retail establishment. It does not include frozen or canned fruits and vegetables or fruit and vegetable juices. Food safety includes work in support of FDA's Foods Program, excluding cosmetics, and Animal Drugs and Feeds Program, excluding medical products.

To determine the dollars and staff years FDA has spent on fresh produce safety, we obtained and analyzed FDA data on food safety and fresh produce safety spending. We used the food safety data to provide context for assessing the relative importance of fresh produce spending and the decisions leading to food safety resource allocations. For both food safety and fresh produce safety spending, some data were obtained directly from FDA databases, and others were estimated by FDA officials.

Food safety spending. We requested information on the dollars and staff years spent on food safety each fiscal year from 2000 through 2007, as well as detailed spending reports showing the dollars and staff years spent on specific food safety activities for each fiscal year from 2005 through 2007, by the five FDA organizations that conducted food safety work—the Center for Food Safety and Applied Nutrition (CFSAN), the Center for Veterinary Medicine (CVM), the Office of Regulatory Affairs (ORA), the National Center for Toxicological Research (NCTR), and the Office of the Commissioner. The Office of the Commissioner's work in support of the agency's food safety efforts is generally considered overhead, except for specific food safety activities such as those performed by FDA's Associate Commissioner for Food and the Office of Crisis Management. We selected fiscal years 2000 through 2007 to update a previous GAO report which detailed food safety spending through fiscal year 1999.[17] Four organizations used time and attendance and workflow management systems to identify hours spent on food safety and then converted the hours into staff years.

- CFSAN used the Resource Reporting System Via Project, its voluntary time reporting system, which captures staff hours spent on one of seven general categories, including four that fall within GAO's definition of food safety,[18] and on specific activities within these categories.
- CVM used the Activity Time Reporting system, its mandatory time reporting system.
- ORA used the Field Accomplishment and Compliance Tracking System (FACTS), its workflow management system, which captures staff time spent on domestic activities, such as inspections or sample analysis, as well as certain import activities, such as sample collections or field exams.
- NCTR used the Project Management System, which contains, among other things, staff and contractor hours downloaded from its NCTR Experiment Activity Tracking System, the center's time and attendance system, as well as contractor hours downloaded from its Task Tracking System.

The Office of the Commissioner calculated its food safety staff years to be 34.6 percent of its total staff years—the percentage it typically uses in FDA's budget justification documents.

To determine the dollars CFSAN spent on food safety, FDA ran their food safety staff years against the agency's Unified Financial Management System. To determine the dollars NCTR spent on food safety, FDA used cost data on projects from the center's Project Management System and the agency's Unified Financial Management System to summarize spending on food safety research projects. To determine dollars ORA spent on food safety, FDA ran its food safety staff years against ORA's actual budget authority dollars from the All Purpose table of the agency's annual budget justification. For the dollars CVM spent on food safety, FDA used CVM's budget authority dollars from the All Purpose table to reflect GAO's definition of food safety because CVM could not isolate medical products spending from other spending. In terms of overhead, the Office of the Commissioner estimated that each year's food safety spending amounted to 34.6 percent of its annual budget authority—the percentage it typically uses in FDA's budget justification documents. Additionally, ORA and CVM included overhead in each activity's spending, while CFSAN and NCTR reported it as a separate expense.

Produce safety spending. We requested information on the dollars and staff years FDA spent each fiscal year from 2000 through 2007. However, due to resource constraints and other limitations within FDA that made it unlikely the agency could provide data or estimates prior to fiscal year 2005, we agreed to narrow our request to fiscal years 2005 through 2007. In addition, FDA was unable to provide its actual spending on fresh produce safety and instead developed estimates, with each of the three organizations involved in produce safety during this period—CFSAN, ORA, and the Office of the Commissioner—employing the same sources and methodology it used to prepare food safety spending data.

- CFSAN could only provide estimates of the dollars and staff years spent on fresh produce safety because its time management system is voluntary and staff are not required to report their time with the specificity needed to distinguish time spent on fresh produce activities from time spent on other food safety activities. Additionally, due to changes in its voluntary time reporting system, CFSAN could provide these estimates only for fiscal years 2006 and 2007. Further, CFSAN's spending estimates reflect only the minimum dollars and staff years spent on fresh produce because staff attributed over half of their food safety staff years to the general category called "food safety," rather than to a specific activity, such as "produce safety."
- ORA was able to track its spending at the fresh produce level for fiscal years 2005 through 2007 because field staff generally enter a commodity code into ORA's work management system as they report time spent on their work activities. However, ORA officials cautioned that their spending data were also estimates because a small number of commodities were coded as miscellaneous and certain activities, such as reviewing imported items for admissibility, do not require staff to report the specific commodity involved.
- The Office of the Commissioner could provide neither data nor estimates of its fresh produce spending. Although some staff—such as the Associate Commissioner for Foods and those in the Office of Crisis Management—work on fresh produce safety, the office's systems do not track spending at the level of specificity needed to, for example, distinguish food-related emergencies from nonfood emergencies or foodborne illness outbreaks involving fresh produce from outbreaks involving other foods.

We reviewed the methodologies FDA used to develop the data and the estimates, interviewed those staff most knowledgeable about the quality and completeness of the estimated data, and conducted tests of the data for errors and missing information. When errors or missing data were found, FDA officials corrected these errors and provided us with revised data. On the basis of this information, we assessed the reliability of the data on food safety spending and the estimates on fresh produce spending and determined that the data and estimates are sufficient and appropriate to support the conclusions reached in this chapter. We then summarized and analyzed the data and estimates to calculate the total dollars and staff years FDA spent each fiscal year on food safety and fresh produce safety. Due to the nature of the estimates, we were unable to assess fresh produce spending over time to determine whether spending increased or draw specific conclusions about the portion of FDA's food safety resources spent on fresh produce.

To determine how FDA has allocated its resources for fresh produce safety, we interviewed staff knowledgeable about the budget and priority-setting processes used by CFSAN, ORA, and the Office of the Commissioner; obtained and analyzed budget and planning documents on agency and center priorities and strategic plans; interviewed current and former FDA officials about the agency's food safety and fresh produce safety work, as well as the obstacles faced in implementing planned work; obtained and analyzed reports from FDA and others on the agency's food safety resources; and interviewed food safety experts and industry representatives for their perspectives on FDA's ability to ensure the safety of fresh produce.

To assess the effectiveness of FDA's actions to oversee fresh produce, we visited produce farms and processing facilities in California's Salinas Valley, where we interviewed growers, processors, and industry representatives, and observed an FDA inspection of a fresh-cut produce facility. We selected the Salinas Valley because it was the source of the 2006 *E. coli* outbreak linked to bagged spinach. Also, we interviewed persons knowledgeable about FDA's oversight activities related to fresh produce, including current and former FDA officials, state food safety officials, industry officials, and food safety experts. We reviewed relevant FDA documents, including policies and procedures, regulations and guidance documents for industry pertaining to fresh produce, and descriptions of research needs and current research projects. We also reviewed guidance documents prepared by industry and obtained information on fresh produce safety standards promoted by other entities. We obtained and analyzed FDA data on domestic inspections and domestic and imported sample analyses from the FACTS database, and FDA data on imported fresh produce trends and activities from the Operational and Administrative System for Import Support database. We assessed the reliability of the data used in this chapter and found it to be sufficiently reliable for the purposes used. We also obtained and analyzed information on state contracts and partnerships from FDA officials and fresh produce-related warning letters from FDA's Web site.

To determine the actions FDA plans to take to enhance fresh produce oversight, we reviewed published information from FDA, including its *Food Protection Plan*, *Food Protection Operations Plan*, and *Federal Register* notices. We also reviewed internal FDA documents and interviewed FDA officials for additional details on published information. To assess the extent to which FDA's planned actions address identified challenges, we reviewed previous GAO reports and recommendations on food safety, food safety publications from the National Academies, and GAO guidance for assessing key elements in agencies' performance plans, including goals, strategies, and resources.[19] We also interviewed and obtained documents from former FDA officials, officials from the U.S. Department of Agriculture and state food safety agencies, academics, and representatives of industry and consumer groups to obtain information related to FDA's planned actions.

We conducted our work from June 2007 to September 2008 in accordance with generally accepted government auditing standards. Those standards require that we plan and perform the audit to obtain sufficient, appropriate evidence to provide a reasonable basis for our findings and conclusions based on our audit objectives. We believe that the evidence obtained provides a reasonable basis for our findings and conclusions based on our audit objectives.

APPENDIX II: FDA FOOD SAFETY SPENDING INFORMATION FOR FISCAL YEARS 2000 THROUGH 2007

This appendix contains information provided by FDA on the dollars and staff years it spent on food safety from fiscal years 2000 through 2007. For this chapter, food safety includes work in support of FDA's Foods Program, excluding cosmetics, and Animal Drugs and Feeds Program, excluding medical products.

Five organizations within FDA provided food safety spending information, including

- the Center for Food Safety and Applied Nutrition (CFSAN), which administers the FDA Foods Program;
- the Center for Veterinary Medicine (CVM), which administers the Animal Drugs and Feeds Program;
- the Office of Regulatory Affairs (ORA), which conducts field work in support of FDA's centers and their programs;
- the National Center for Toxicological Research, which conducts scientific research and provides technical advice in support of FDA's centers; and
- the Office of the Commissioner, which provides direction and administrative services in managing FDA's food safety efforts.

The information provided by each organization varies in the level of detail and the time period covered.

Table 5. Summary of FDA Food Safety Spending, by Organization, Fiscal Years 2000 through 2007

Dollars in millions

	2000		2001		2002	
Organization	**Dollars**	**Staff years**	**Dollars**	**Staff Years**	**Dollars**	**Staff years**
Center for Food Safety and Applied Nutrition (CFSAN)	$120.9	793	$122.2	842	$140.3	894
ORA fieldwork in support of CFSAN	154.2	1,545	160.5	1,554	248.9	1,799
Center for Veterinary Medicine (CVM)	36.5	271	48.4	290	55.7	323
ORA fieldwork in support of CVM	12.5	128	14.8	144	28.4	235
National Center for Toxicological Research[b]	3.9	24	3.0	19	8.0	27
Office of the Commissioner[c]	23.0	226	23.5	233	28.4	231
Total	**$351.1**	**2,988**	**$372.4**	**3,082**	**$509.7**	**3,509**

Table 5. (Continued)

Fiscal year									
2003		2004		2005		2006		2007	
Dollars	Staff years	Dollars	Staff years	Dollars	Staff years	Dollars	Staff years	Dollars	Staff years
$143.8	921	$140.9	881	$148.4	786	$150.3	816[a]	$157.1	744
257.9	2,203	260.8	2,157	281.4	2,045	283.5	1,950	296.1	1,793
57.1	341	54.5	346	55.4	330	54.8	321	58.4	318
29.0	242	27.5	234	33.4	229	33.0	206	34.6	208
7.9	36	7.4	36	6.5	34	10.3	36	11.5	37
29.1	226	31.2	199	30.2	176	29.6	184	31.6	181
$524.8	**3,969**	**$522.4**	**3,853**	**$555.3**	**3,599**	**$561.6**	**3,513**	**$589.1**	**3,281**

Source: GAO analysis of FDA data.

Notes: Numbers may not add due to rounding.

[a]CFSAN's staff years spent in fiscal year 2006 exceeded its budget authority due to the inclusion of overtime and credit hours.

[b]Spending includes dollars received from other government agencies and external sources for collaborative efforts.

[c]The Office of the Commissioner developed an overhead figure of 34.6 percent of its budget authority to reflect dollars and staff years spent on food safety.

Table 6. CFSAN Food Safety Spending, by Category, Fiscal Years 2006 and 2007

Dollars in millions				
	Fiscal year			
	2006		2007	
Category	**Dollars**	**Staff years**	**Dollars**	**Staff years**
Food defense	$20.6	113	$22.5	89
Food safety[a]	102.3	550	108.4	531
Nutrition	4.7	36	4.3	26
Dietary supplements	5.1	26	5.7	26
Overhead[b]	17.6	91	16.2	71
Total	**$150.3**	**816**	**$157.1**	**744**

Source: FDA.

Notes: Numbers may not add due to rounding. CFSAN was not able to provide data at this level of detail prior to fiscal year 2006. Also, CFSAN's total food safety spending includes $18 million in fiscal year 2006 and $28 million in fiscal year 2007 in work conducted through major contracts—each totaling at least $200,000—with other government agencies, as well as external organizations and individuals.

[a]Food safety refers to CFSAN's traditional definition of food safety—the unintentional contamination of food.

[b]Overhead represents CFSAN's prorated share of the agency's overhead, such as human resources, information technology, management services, and telephone expenses.

Table 7. ORA Food Safety Spending in Support of CFSAN, by Category, Fiscal Years 2005 through 2007

Dollars in millions						
	Fiscal year					
	2005		2006		2007	
Category	Dollars	Staff years	Dollars	Staff years	Dollars	Staff years
Chemical safety of foods						
Pesticides and industrial chemicals in foods (domestic and import)	$23.2	181	$19.2	139	$21.3	136
Mycotoxins in foods (domestic and import)	5.6	44	5.7	41	5.9	38
Food and color additives (import)	4.7	37	5.1	37	6.1	39
Toxic elements in foods (domestic and import)	4.8	37	4.0	29	4.2	27
Chemotherapeutics in seafood	3.3	26	4.1	29	3.9	25
Field assignment for chemical contaminants	2.8	22	1.7	12	1.3	8
Forensic analysis[a]	2.1	16	2.0	15	1.4	9
Toxic elements in foodwares (domestic and import)	2.0	15	1.8	13	1.6	10
Radionuclides in foods	0.2	1	0.3	2	0.7	4
Pet food recall (human foods)	0.0	0	0.0	0	2.4	15
Microbiological safety of foods						
Cheese program (domestic and import)	7.2	56	7.1	51	7.5	48
Retail food protection program (general)	5.7	44	6.5	47	7.1	46
Interstate travel program	5.8	45	5.7	41	6.0	38
Domestic acidified and low acid canned food program	4.2	33	5.2	38	5.2	33
Microbiological assignment (domestic and import)	4.0	31	4.9	35	2.3	15
Interstate milk shippers program	4.3	33	4.6	34	6.1	39
Import acidified and low acid canned food program	3.5	27	3.9	28	4.1	26
Molluscan shellfish evaluation program	3.2	25	3.6	26	4.1	26
Audits of state food contract inspections	1.2	10	2.0	14	2.5	16
Natural disasters and emergencies	0.9	7	4.7	34	0.1	<1
Nutrient quality and food labeling						
Nutrition Labeling and Education Act, nutrient sample analysis, and general food labeling	7.7	60	7.8	56	8.5	54
Dietary supplements program	3.0	23	3.0	22	2.5	16
Medical foods (domestic and import)	1.5	12	0.9	7	0.9	6
Infant formula	1.1	9	1.8	13	1.5	10
Health fraud (foods)	0.6	5	0.7	5	0.9	6
Food economics and standards	0.4	3	0.4	3	0.2	1
Cross-cutting						
Imported foods (general)	58.7	458	64.2	464	72.6	465
Domestic food safety program (general)	23.9	186	24.9	180	22.3	143
Domestic seafood hazard analysis and critical control point program	23.2	181	20.9	151	20.6	132
Imported seafood hazard analysis and critical control point program	22.0	172	22.3	161	21.7	139
Total diet studies	6.8	53	8.9	64	8.6	55
Counterterrorism activities	6.8	53	2.8	21	2.3	15

Table 7. (Continued)

Dollars in millions						
	Fiscal year					
	2005		2006		2007	
Category	Dollars	Staff years	Dollars	Staff years	Dollars	Staff years
Consumer complaints	4.8	38	5.0	36	4.3	28
Methods validation and development	4.1	32	4.4	32	4.7	30
Juice hazard analysis and critical control point program	3.1	24	3.9	28	4.2	27
Short-term assignments initiated by CFSAN headquarters, ORA headquarters, or ORA field offices	2.0	16	2.0	14	1.5	10
Foreign inspections and technical assistance	1.7	13	1.6	12	1.4	9
Emergency response to foodborne outbreaks and illnesses	0.9	7	1.2	8	7.1	46
Miscellaneous other food safety work	0.6	5	0.7	5	0.3	2
Criminal investigation activities	0.6	5	0.6	4	0.4	2
Subtotal	**262.3**	**2,045**	**270.0**	**1,950**	**280.3**	**1,793**
State contracts and grants	19.1	0	13.5	0	15.8	0
Total	**$281.4**	**2,045**	**$283.5**	**1,950**	**$296.1**	**1,793**

Source: GAO analysis of FDA data.

Notes: Numbers may not add due to rounding. ORA included overhead for each activity as part of the activity's total dollars and staff years spent.

[a]Spending on other forensic analysis work is included in activities where work occurred.

Table 8. Center for Veterinary Medicine Spending on Food Safety, by Category, Fiscal Years 2000 through 2007

Dollars in millions						
	2000		2001		2002	
Category	Dollars	Staff years	Dollars	Staff Years	Dollars	Staff years
Pre-market						
Review	$14.1	126	$19.2	141	$21.3	149
Applied research	3.6	25	4.1	25	4.3	26
Outreach/coordination	0.9	12	0.9	12	1.0	12
Post-market						
Outreach coordination compliance	11.8	76	17.4	79	21.2	100
Applied research	6.1	32	6.8	33	7.9	36
Total	**$36.5**	**271**	**$48.4**	**290**	**$55.7**	**323**

Table 8. (Continued)

Fiscal Year									
2003		2004		2005		2006		2007	
Dollars	Staff years	Dollars	Staff years	Dollars	Staff years	Dollars	Staff years	Dollars	Staff years
$21.8	158	$21.4	176	$21.1	152	$25.2	145	$27.7	149
4.4	26	4.1	26	4.1	27	1.7	9	1.8	14
1.0	13	0.8	12	0.8	12	0.5	4	0.5	4
21.9	108	20.6	97	21.7	101	22.0	132	22.2	109
8.1	36	7.6	35	7.6	38	5.3	31	6.2	42
$57.1	**341**	**$54.5**	**346**	**$55.4**	**330**	**$54.8**	**321**	**$58.4**	**318**

Source: FDA.

Note: Numbers may not add due to rounding.

Table 9. ORA Food Safety Spending in Support of CVM, by Category, Fiscal Years 2005 through 2007

Dollars in thousands	**Fiscal year**					
Category	**2005**		**2006**		**2007**	
	Dollars	**Staff years**	**Dollars**	**Staff years**	**Dollars**	**Staff years**
Drug processing and new animal drug inspection program	$2,683	21	$2,356	17	$2,558	18
Feed contaminants program	4,154	33	3,564	26	3,440	25
Feed manufacturing inspection program	2,769	22	2,710	20	1,763	13
Illegal residues in meat and poultry	3,087	25	4,065	30	2,343	17
Ruminant feed ban rule and BSE inspection program	11,510	91	9,720	72	6,019	43
Methods validation and development	952	8	766	6	688	5
Forensic analysis[a]	0	0	0	0	0	0
Audits of state contract inspections	0	0	295	2	516	4
Import entry review and refused entry tracking	1,904	15	2,121	16	1,333	10
Illegal sales, compounding, counterfeit animal drugs	404	3	265	2	64	1
Animal feed short-term assignments, ORA or field directed	663	5	1,001	7	537	4
Animal feed consumer complaints	202	2	265	2	2,107	15
Animal feed criminal investigation activities	144	1	501	4	107	1
Miscellaneous other activities	375	3	324	2	322	2
Pet food recall activities	0	0	0	0	7,288	52
Subtotal	**28,848**	**229**	**27,952**	**206**	**29,086**	**208**
State contracts and grants	4,520	0	5,066	0	5,482	0
Total	**$33,368**	**229**	**$33,018**	**206**	**$34,568**	**208**

Source: FDA.

Notes: Numbers may not add due to rounding.

[a]Forensic analysis work is included in activities where work occurred.

Table 10. National Center for Toxicological Research Spending on Food Safety, by Category, Fiscal Years 2000 through 2007

Dollars in thousands

	2000		2001		2002	
Category	Dollars	Staff years	Dollars	Staff years	Dollars	Staff years
Food safety	$1,015	12	$325	3	$694	7
Antimicrobial resistance	785	7	791	10	1,172	12
Bioterrorism	141	1	65	1	3,178	3
Dietary supplements	560	4	786	5	925	5
Overhead	1,432	[a]	1,010	[a]	2,002	[a]
Total	**$3,933**	**24**	**$2,977**	**19**	**$7,971**	**27**

	Fiscal year									
	2003		2004		2005		2006		2007	
Category	Dollars	Staff years	Dollars	Staff years	Dollars	Staff years	Dollars	Staff years	Dollars	Staff years
Food safety	$293	8	$1,832	13	$1,463	10	$3,874	10	$3,718	7
Antimicrobial resistance	1,114	17	1,298	16	1,404	17	1,510	15	1,682	17
Bioterrorism	2,842	6	424	2	113	1	551	5	797	7
Dietary supplements	1,126	5	1,023	5	1,530	6	1,774	6	2,439	6
Overhead	2,518	[a]	2,832	[a]	2,033	[a]	2,599	[a]	2,848	[a]
Total	**$7,893**	**36**	**$7,409**	**36**	**$6,542**	**34**	**$10,308**	**36**	**$11,483**	**37**

Source: FDA.

Notes: Numbers may not add due to rounding. Spending includes dollars received from other government agencies and external sources for collaborative efforts.

[a]Information not provided.

APPENDIX III: COMMENTS FROM THE DEPARTMENT OF HEALTH AND HUMAN SERVICES

DEPARTMENT OF HEALTH & HUMAN SERVICES OFFICE OF THE SECRETARY

Assistant Secretary for Legislation
Washington, DC 20201

SEP 10 2008

Lisa Shams
Director, Natural Resources and Environment
U.S. Government Accountability Office
441 G Street N.W.
Washington, DC 20548

Dear Ms. Shams:

Thank you for the opportunity to review and comment on the U.S. Government Accountability Office's (GAO) report entitled: "Food Safety: Improvements Needed in FDA Oversight of Fresh Produce" (GAO 08-1047). This is to notify you that we have reviewed the subject final report, and we have no additional comments to offer since the publication of the GAO draft report.

The Department appreciates the effort that went into this report.

Sincerely,

Jennifer R. Luong
for Vincent J. Ventimiglia, Jr.
Assistant Secretary for Legislation

Attachment

Note: GAO comments supplementing those in the report text appear at the end of this appendix.

COMMENTS OF THE DEPPARTMENT OF HEALTH AND HUMAN SERVICES (HHS) ON THE U.S. GOVERNMENT ACCOUNTABILITY OFFICE'S (GAO) DRAFT REPORT ENTITLED: "FOOD SAFETY — IMPROVEMENTS NEEDED IN FDA OVERSIGHT OF FRESH PRODUCE" (GAO-08-1047)

The Food and Drug Administration (FDA) appreciates the opportunity to review and comment on the Government Accountability Office's (GAO) draft report, *Food Safety: Improvements Needed in FDA Oversight of Fresh Produce* (GAO-08-1047). The Agency will offer both general and specific comments. Of significant importance to FDA is GAO's recognition of the Food Protection Plan (FPP) as a sound framework for advancing food safety and food defense[20] and that the GAO utilized the plan's framework as an organizing feature in this chapter.

In November 2007 FDA released the FPP which provides a national strategy to identify and counter potential hazards in both domestic and imported food. The FPP was developed to address the changes and challenges in protecting the U.S. food supply. The FPP places a greater emphasis on prevention by building in safety at the source of food production, regardless of whether food is produced domestically or outside the U.S. Prevention is facilitated through targeted risk-based intervention strategies that involve risk-based inspections and testing. The FPP also seeks to develop faster response and enhanced communication with stakeholders during food related events. While the focus of the current GAO report is on fresh produce safety, the elements of the FPP encompass all foods for humans and animals Many of the specific actions in the FPP to improve prevention, intervention and response are being applied to fresh produce. Achieving the food safety enhancements identified by the FPP will require the involvement of all our food safety partners – Federal, state, local, tribal, and foreign governments; industry; academia; consumers; and Congress.

GAO's approach, using the FPP framework, helps solidify a common food vision toward protecting the food supply and provides a framework for discussion of the issues such that all stakeholders can work uniformly to advance food protection.

In this draft report, GAO makes 7 recommendations to FDA, including 4 recommendations to enhance its oversight of fresh produce safety, such as updating its good agricultural practices guidance, 2 recommendations to seek authority from Congress to make explicit FDA's authority to adopt preventive controls for high-risk foods and to provide enhanced access to firm records during food related emergencies, and one recommendation to provide specific information to Congress and to the public on the strategies and resources for implementing the FPP. As part of the FPP, FDA made 10 legislative proposals and while none of them are exclusively focused on fresh produce they are all important in ensuring that the food supply in the U.S. is protected. FDA's response to the specific GAO recommendations follows:

Recommendations to Enhance Oversight of Fresh Produce Safety

First GAO recommendation – that FDA develop a plan for identifying research priorities and facilitating research related to fresh produce.

FDA agrees with this recommendation. Both CFSAN and the Agency, as part of the Food Protection Plan, are currently developing strategic plans for research which will include but not be limited to fresh produce related research. The Agency's regulatory food safety research needs are far broader than fresh produce; therefore, fresh produce research needs must, out of necessity, compete for limited resources. CFSAN 's plan will identify regulatory research priorities that can be addressed through (1) intramural research (2) extramural research leveraged through our Centers of Excellence or (3) through a competitive extramural research program. The plan also will identify future research needs that currently cannot be addressed due to resource (funding and expertise) limitations.

> The 2004 Produce Safety Action Plan (http://www.cfsan.fda.govi–dms/prodpla2.html) identified broad research needs relevant to the contamination of fresh produce. CFSAN has systematically worked with our stakeholders and other research organizations (USDA, ARS and CSREES; WIFSS; NCFST and JIFSAN[21]) to leverage research resources and identify more specific research needs associated with fresh produce. These activities included participating in workshops such as the Tomato Safety Research Needs Workshop (http:,/www.nfsan.umd.edu/tomato wkp2007.htm); the International Lettuce and Leafy Greens Food Safety Research Conference (http://www.unitedfresh.orginewsviewsileafy greens food safety research); and identifying appropriate research focus areas for the USDA's National Research Initiative (http: 'wWW.csrees.usda.govitundinginriinri.html) and the National Integrated Food Safety Initiative (http:,/www.csrees.usda.goy/fundingirfasifood _safety.html) programs.

Second GAO recommendation --that FDA identify approaches for obtaining testing and other information from industry members to supplement its research agenda

FDA agrees with the recommendation to identify approaches for obtaining testing and other information from industry members. FDA would like to note that the data and information from industry would further inform, not supplement, FDA's research agenda. Additionally, the data obtained would be used in Agency risk assessments associated with fresh produce. The Agency is aware of efforts by some produce industry associations and UC Davis's Center for Produce Safety regarding how they might serve as repositories of such data and information for the purpose of sharing with others including FDA.

Third GAO recommendation — that FDA update its good agricultural practices guidance for fresh produce to incorporate knowledge about safe growing practices

FDA agrees with this recommendation. The Agency recognized when it issued the good agricultural practices guidance in 1998 that it would need to be updated as new information and technological advances expand the understanding of those factors associated with identifying and reducing microbial food safety hazards. In the 10 years since the guidance was released, many changes have occurred in the produce industry, and a great deal of new knowledge and information has become available. In addition, the Agency now has ten years experience in implementing the guidance, and observing how and the extent to which it has been implemented.

Although at the time GAO interviewed FDA, drafting of revised guidance had not started, many of the actions necessary to update the guidance had been completed or were being done. For example, FDA held public hearings on produce safety on March 13 and April 20, 2007, where the Agency solicited input from stakeholders on risk factors associated with microbial contamination of produce. In addition, the Agency has implemented the leafy greens and tomato initiatives, whose goals include gaining knowledge on environmental factors that lead to contamination of produce, which will inform future policy or guidance.

More recently, the Agency published a notice in the Federal Register on September 2, 2008, requesting comments and scientific data and information that may assist the Agency to improve the good agricultural practices guidance. Specifically, FDA is seeking information about (1) current agricultural practices and conditions used to grow, harvest, pack, cool, and transport fresh produce; (2) risk factors for contamination of fresh produce associated with these practices; and (3) possible measures that FDA could implement that would enhance the safety of fresh produce. The Agency will update the guidance after evaluating comments it receives together with all the data and information it has collected through the other efforts mentioned above.

Fourth GAO Recommendation -- that FDA update its current good manufacturing practice regulations for food to incorporate new knowledge about the food industry and safe manufacturing, processing, and holding practices

FDA agrees with this recommendation. Beginning in late 2002, the Center for Food Safety and Applied Nutrition formed a Food Current Good Manufacturing Practice (CGMP) Modernization Working Group. The objective of the group was to examine the general food CGMP regulation in 21 CFR Part 110 (hereinafter CGMP regulation) and determine whether the regulation was in need of modernization. Also, the group was specifically tasked to focus on risk-based preventive controls, i.e. those that would have the greatest impact on ensuring food safety. The working group concluded that there have been changes in both the food industry and in the science of food safety that indicate a need for modernization.

In 2003, the working group initiated research programs to identify those areas where GMP-type controls could have the greatest impact on ensuring food safety. In 2004, the working group presented its preliminary findings from this research and engaged the public in three public meetings held across the country and through a Federal Register notice calling for comments on food CGMP modernization. This chapter, found at http:dwww.ctsanicla. gov'~dmsicgmps.html, summarizes the public comments and details the working group's key findings.

The areas that present opportunities for modernization are training, food allergens, Listeria monocytogenes control, sanitation procedures, and application of certain CGMPs to agricultural operations, records access, and temperature control.

Recommendations to Seek Authority from Congress

Fifth and sixth GAO recommendations – that the Commissioner of FDA seek authority from the Congress to make explicit FDA's authority to adopt preventive controls for high-risk goods, and provide FDA enhanced access to firm records during food-related emergencies.

FDA has already sought authority from Congress for these and additional legislative proposals. These requests are highlighted in the Food Protection Plan, as well as Congressional testimony. To date, Congress has not acted on the legislative needs outlined in the Food Protection Plan. As a package, these proposals will not only enhance food safety overall but will specifically contribute to enhanced produce safety. For example, additional authority to authorize FDA to accredit highly-qualified third parties for voluntary food inspection is outlined in the Food Protection Plan, but not referenced above in the GAO recommendation. This authority would allow FDA to leverage its inspectional capabilities without relinquishing any of its regulatory or enforcement authority. As noted in the GAO report, a voluntary third party accreditation program focused on fresh produce could provide several benefits while increasing oversight of a high risk food category.

FDA refers GAO to its own outstanding recommendation that Congress enact comprehensive, uniform, and risk-based food safety legislation.[22] FDA supports GAO's recommendation for Congress to act on food safety legislation and proposes that the recommendation above be redirected to Congress for action.

Recommendation to Provide Information to Congress

Seventh GAO recommendation — that the Commissioner of FDA provide specific information to the Congress on the strategies and resources for implementing the Food Protection Plan

FDA agrees with the need for transparency and accountability in implementing the Food Protection Plan. FDA has provided extensive detail to Congress on the implementation plan – including resources – for the Food Protection Plan. FDA intends to continue to be as transparent as possible in its implementation of the FPP in order to enable FDA, Congress, and the public to track progress on the Food Protection Plan.

The following are GAO's comments on FDA's written comments provided by the Department of Health and Human Service's Assistant Secretary for Legislation in a letter dated September 10, 2008.

GAO Comments

1. FDA did not agree with our recommendations to seek authority from the Congress to make explicit FDA's authority to adopt preventive controls for high-risk foods and to provide FDA enhanced access to firm records during food-related emergencies. FDA

believes that the agency has already sought authority from the Congress for these and additional legislative authorities. Specifically, FDA said that the agency has sought authority by outlining these legislative needs in the *Food Protection Plan* and highlighting them in congressional testimony. We do not view these actions as seeking authority because FDA has not drafted legislative language or formally submitted a legislative proposal to the Congress, nor has it worked directly with the Congress to help initiate these authorities. FDA proposed that these recommendations be redirected to the Congress for action and refers to a prior recommendation we made that the Congress enact comprehensive, uniform, and risk-based food safety legislation. We reiterated this recommendation to the Congress most recently in our High-Risk Series[23] as one action that can help address the fragmented federal oversight of food safety and integrate the myriad food safety programs. Our recommendations to FDA are intended to enhance FDA's authority to oversee fresh produce for which FDA has primary responsibility.

END NOTES

[1] GAO, *High-Risk Series: An Update, GAO-07-310* (Washington, D.C.: January 2007).

[2] FDA Science Board, Subcommittee on Science and Technology, FDA *Science and Mission at Risk* (Washington, D.C., November 2007).

[3] Department of Health and Human Services, U.S. Food and Drug Administration, *Food Protection Plan: An Integrated Strategy for Protecting the Nation's Food Supply* (Washington, D.C., November 2007).

[4] GAO, *Federal Oversight of Food Safety: FDA's Food Protection Plan Proposes Positive First Steps, but Capacity to Carry Them Out Is Critical,* GAO-08-435T (Washington, D.C.: Jan. 29, 2008) and GAO, *Federal Oversight of Food Safety: FDA Has Provided Few Details on the Resources and Strategies Needed to Implement its Food Protection Plan*, GAO-08-909T (Washington, D.C.: June 12, 2008).

[5] GAO, *Food Safety: Overview of Federal and State Expenditures*, GAO-01-177 (Washington, D.C.: Feb. 20, 2001).

[6] GAO, *Agencies' Annual Performance Plans under the Results Act: An Assessment Guide to Facilitate Congressional Decisionmaking*, GAO/GGD/AIMD-10.1.18 (Washington, D.C.: February 1998).

[7] According to FDA, these data do not include (1) outbreaks and illnesses where the point of contamination is the retail food setting or home, (2) illnesses transmitted from person-to-person, and (3) illnesses or deaths that may have occurred but were not captured by the outbreak reporting process.

[8] According to ORA, it hired an additional 597 people in fiscal year 2002 to help conduct work related to the Bioterrorism Act.

[9] GAO-08-909T.

[10] These centers are the National Center for Food Safety and Technology at the Illinois Institute for Technology, the Joint Institute for Food Safety and Applied Nutrition at the University of Maryland, and the National Center for Natural Products Research at the University of Mississippi. FDA is in the process of developing a fourth center, the Western Center for Food Safety, at the University of California, Davis.

[11] According to FDA's Office of Chief Counsel, formal enforcement actions available to FDA include initiating a seizure of an adulterated product, obtaining an injunction to stop a company from engaging in a certain behavior, or referring a firm for criminal prosecution. Warning letters are intended to prompt voluntary corrections for violations of regulatory significance. Additionally, FDA may issue a less formal, untitled letter or request a regulatory meeting for less severe violations.

[12] According to FDA, an entry line means each portion of an import shipment that is listed as a separate item on an import document. Items in an import entry having different tariff descriptions must be listed separately.

[13] GAO, *Food Safety: Changes Needed to Minimize Unsafe Chemicals in Food*, GAO/RCED-94-192 (Washington, D.C.: Sept. 26, 1994).

[14] GAO, *Medical Devices: Challenges for FDA in Conducting Manufacturer Inspections*, GAO-08-428T (Washington, D.C.: Jan. 29, 2008).

[15] GAO, *Federal Oversight of Food Safety: High-Risk Designation Can Bring Attention to Limitations in the Federal Government's Food Recall Programs*, GAO-07-785T (Washington, D.C.: Apr. 24, 2007) and GAO,

Food Safety: USDA and FDA Need to Better Ensure Prompt and Complete Recalls of Potentially Unsafe Food, GAO-05-51 (Washington, D.C.: Oct. 6, 2004).

[16] GAO-07-310.

[17] GAO, *Food Safety: Overview of Federal and State Expenditures*, GAO-01-177 (Washington, D.C.: Feb. 20, 2001).

[18] These four categories are food safety, food defense, dietary supplements, and nutrition and labeling.

[19] GAO, *Agencies' Annual Performance Plans under the Results Act: An Assessment Guide to Facilitate Congressional Decisionmaking*, GAO/GGD/AIMD-10.1.18 (Washington, D.C.: February 1998).

[20] GAO, *Federal Oversight of Food Safety: FDA Has Provided Few Details on the Resources and Strategies Needed to Implement its Food Protection Plan*. GAO-08-909T (Washington, D.C.: July 12, 2008).

[21] Acronyms: United States Department of Agriculture (USDA), Agricultural Research Service and Cooperative State Research, Education, and Education Service (ARS and CSREES), Western Institute for Food Safety and Security (WIFSS), National Center for Food Safety and Technology (NCFST), and Joint Institute for Food Safety and Nutrition (JIFSAN).

[22] GAO Testimony Before the Subcommittee on Oversight and Investigations, Committee on Energy and Commerce, House of Representatives, *Federal Oversight of Food Safety: High-Risk Designation Can Bring Attention to the Limitations in the Government's Food Recall Programs*, April 24, 2007

[23] GAO, *High-Risk Series: An Update,* GAO-07-310 (Washington, D.C.: January 2007).

In: Food Safety, Fresh Produce...
Editor: Russell P. Barton
ISBN: 978-1-60692-788-5

Chapter 2

THE FEDERAL FOOD SAFETY SYSTEM: A PRIMER

Geoffrey S. Becker and Donna V. Porter

SUMMARY

Numerous federal, state, and local agencies share responsibilities for regulating the safety of the U.S. food supply, which many experts say is among the safest in the world. Nevertheless, critics view this system as lacking the organization and resources to adequately combat foodborne illness — as evidenced by a series of widely publicized food safety problems, including concerns about adulterated food and food ingredient imports, and illnesses linked to various types of fresh produce and to meat and poultry products. Numerous bills addressing various aspects of food safety have been introduced into the first session the 110th Congress, including proposals to reorganize oversight authorities, increase funding, tighten regulation of imported foods, and establish new authorities for enforcement, recall, and notification regarding adulterated foods, among others. Portions of several proposals have become law; more sweeping changes in the federal food safety system are still under consideration in 2008.

BACKGROUND[1]

Americans spend more than $1 trillion on food each year, nearly half of it in restaurants, schools, and other places outside the home.[2] Federal laws give food manufacturers, distributors, and retailers the basic responsibility for assuring that foods are wholesome, safe, and handled under sanitary conditions. A number of federal agencies, cooperating with state, local, and international entities, play a major role in regulating food quality and safety under these laws.

The combined efforts of the food industry and the regulatory agencies often are credited with making the U.S. food supply among the safest in the world. Nonetheless, public health officials estimate that each year 76 million people become sick, 325,000 are hospitalized, and 5,000 die from foodborne illnesses caused by contamination from any one of a number of microbial pathogens.[3] At issue is whether the current system has the resources and structural organization to protect consumers from these dangers. Also at issue is whether federal food safety laws, first enacted in the early 1900s, have kept pace with the significant changes that have occurred in the food production, processing, and marketing sectors since then.

THE AGENCIES AND THEIR ROLES

The Government Accountability Office (GAO) has identified 15 federal agencies collectively administering at least 30 laws related to food safety. The Food and Drug Administration (FDA), which is part of the U.S. Department of Health and Human Services (HHS), and the Food Safety and Inspection Service (FSIS), which is part of the U.S. Department of Agriculture (USDA), together comprise the majority of both the total funding and the total staffing of the government's food regulatory system.[4] FSIS's FY2007 budget was approximately $930 million in appropriated funds plus another estimated $135 million in industry-paid user fees. FDA's budget for foods was $457 million in FY2007, virtually all of it appropriated.[5]

Among other agencies with smaller but still significant shares of the food safety portfolio are the National Marine Fisheries Service (NMFS), which is part of the U.S. Department of Commerce (DOC), the Environmental Protection Agency (EPA), and the Centers for Disease Control and Prevention (CDC) in HHS.

Food and Drug Administration

The FDA is responsible for ensuring that all domestic and imported food products — except for most meats and poultry — are safe, nutritious, wholesome, and accurately labeled. Examples of FDA-regulated foods are produce, dairy products, seafood, and processed foods. FDA has jurisdiction over meats from animals or birds that are not under the regulatory jurisdiction of FSIS. FDA shares responsibility for the safety of eggs with FSIS. FDA has jurisdiction over establishments that sell or serve eggs or use them as an ingredient in their products. FDA is also responsible for ensuring that most seafood products do not endanger public health (FSIS is to begin inspecting farmed catfish products under a 2008 farm bill provision). The primary statutes governing FDA's activities are the Federal Food, Drug, and Cosmetic Act, as amended (21 U.S.C. 301 et seq.); the Public Health Service Act, as amended (42 U.S.C. 201 et seq.); and the Egg Products Inspection Act, as amended (21 U.S.C. 1031 et seq.).

FDA's food inspection force numbers more than 1,900 in field offices throughout the United States, plus nearly 900 in the Washington, DC, area. FDA regulates food manufacturers' safety practices by relying on companies' self-interest in producing safe products, and by working with the industry to improve production practices. About 57,000 food manufacturers are subject to periodic FDA inspection for regulatory compliance. According to GAO, unannounced compliance inspections of individual establishments by FDA officials occur roughly once every five years. FDA relies on notifications from within the industry or from other federal or state inspection personnel, as well as other sources, to alert it to situations calling for increased inspection.

In the Washington, DC, area, two FDA offices are the focal point for food safety-related activities. The Center for Food Safety and Applied Nutrition (CFSAN) is responsible for (1) conducting and supporting food safety research; (2) developing and overseeing enforcement of food safety and quality regulations; (3) coordinating and evaluating FDA's food surveillance and compliance programs; (4) coordinating and evaluating cooperating states' food safety activities; and (5) developing and disseminating food safety and regulatory information to consumers and industry. FDA's Center for Veterinary Medicine (CVM) is responsible for ensuring that all animal drugs, feeds (including pet foods), and veterinary devices are safe for animals, are properly labeled, and produce no human health hazards when used in food-producing animals.

The FDA also cooperates with over 400 state agencies across the nation that carry out a wide range of food safety regulatory activities. However, the state agencies are primarily responsible for actual inspection. FDA works with the states to set the safety standards for food establishments and commodities and evaluates the states' performance in upholding such standards as well as any federal standards that may apply. FDA also contracts with states to use their food safety agency personnel to carry out certain field inspections in support of FDA's own statutory responsibilities.

Food Safety and Inspection Service

FSIS regulates the safety, wholesomeness, and proper labeling of most domestic and imported meat and poultry and their products sold for human consumption. Under the Federal Meat Inspection Act of 1906, as amended (21 U.S.C. 601 et seq.), FSIS inspects all cattle, sheep, swine, goats, and equines during slaughtering and processing. Under the Poultry Products Inspection Act of 1957, as amended (21 U.S.C. 451 et seq.), FSIS is required to inspect "any domesticated bird" being processed for human consumption; however, USDA regulations implementing this law limit the definition of domesticated birds to chickens, turkeys, ducks, geese, ratites (emus, ostriches, and rheas), and guineas. FDA has jurisdiction over exotic and alternative meats not inspected by FSIS, and shares the responsibility for egg safety with FSIS. The latter is responsible for the safety of liquid, frozen, and dried egg products, domestic and imported, and for the safe use or disposition of damaged and dirty eggs under the Egg Products Inspection Act, as amended (21 U.S.C. 1031 et seq.).

FSIS staff numbers around 9,400; roughly 8,000 of them, including about 1,000 veterinarians, are in about 6,300 meat slaughtering and/or processing plants nationwide. FSIS personnel inspect all meat and poultry animals at slaughter on a continuous basis, and at least one federal inspector is on the line during all hours the plant is operating. Processing inspection does not require an FSIS inspector to remain constantly on the production line or to inspect every item. Instead, inspectors are on site daily to monitor the plant's adherence to the standards for sanitary conditions, ingredient levels, and packaging, and to conduct statistical sampling and testing of products. Because all plants are visited daily, processing inspection also is considered to be continuous.

FSIS also is responsible for certifying that foreign meat and poultry plants are operating under an inspection system equivalent to the U.S. system before they can export their product to the United States. FSIS inspectors located at U.S. ports of entry carry out a statistical sampling program to verify the safety of imported meats from cattle, sheep, swine, goats, and equines and imported poultry meat from chickens, turkeys, ducks, geese, quail, ratites, and guineas before they are released into domestic commerce. FDA is responsible for ensuring the safety of imported meat from any other species.

Twenty-seven states operate their own meat and/or poultry inspection programs. FSIS is statutorily responsible for ensuring that the states' programs are at least equal to the federal program. Plants processing meat and poultry under state inspection can market their products only within the state (although the 2008 farm bill will permit interstate shipment in some cases). If a state chooses to discontinue its own inspection program, or if FSIS determines that it does not meet the agency's equivalency standards, FSIS must assume the responsibility for inspection if the formerly state-inspected plants are to remain in operation. FSIS also has cooperative agreements with more than two dozen states under which state inspection personnel are authorized to carry out federal inspection in meat and/or poultry plants. Products from these plants may travel in interstate commerce.

Centers for Disease Control and Prevention (HHS)

CDC is responsible for (1) monitoring, identifying, and investigating foodborne disease problems to determine the contributing factors; (2) working with FDA, FSIS, NMFS, state and local public health departments, universities, and industry to develop control methods; and (3) evaluating the effect of control methods. In 1995, CDC launched "FoodNet," a collaborative project with the FDA and USDA to improve data collection on foodborne illness outbreaks. FoodNet includes active surveillance of clinical microbiology laboratories to obtain a more accurate accounting of positive test results for foodborne illness; a physician survey to determine testing and laboratory practices; population surveys to identify illnesses not reported to doctors; and research studies to obtain new and more precise information about which food items or other exposures may cause diseases. FoodNet data allows CDC to have a clearer picture of the incidence and causes of foodborne illness and to establish baseline data against which to measure the success of changes in food safety programs. The Public Health Service Act provides legislative authority for CDC's food safety related activities.

National Marine Fisheries Service (DOC)

Although the FDA is the primary agency responsible for ensuring the safety, wholesomeness and proper labeling of domestic and imported seafood products, NMFS conducts, on a fee-for-service basis, a voluntary seafood inspection and grading program that focuses on marketing and quality attributes of U.S. fish and shellfish. The primary legislative authority for NMFS's inspection program is the Agricultural Marketing Act of 1946, as amended (7 U.S.C. 1621 et seq.). NMFS has approximately 160 seafood safety and quality inspectors, and inspection services are funded with user fees.

Environmental Protection Agency

EPA has the statutory responsibility for ensuring that the chemicals used on food crops do not endanger public health. EPA's Office of Pesticide Programs is the part of the agency that (1) registers new pesticides and determines residue levels for regulatory purposes; (2) performs special reviews of pesticides of concern; (3) reviews and evaluates all the health data on pesticides; (4) reviews data on pesticides' effects on the environment and on other species; (5) analyzes the costs and benefits of pesticide use; and (6) interacts with EPA regional offices, state regulatory counterparts, other federal agencies involved in food safety, the public, and others to keep them informed of EPA regulatory actions. The Federal Insecticide, Fungicide, and Rodenticide Act, as amended (7 U.S.C. 136 et seq.), and the Federal Food, Drug, and Cosmetic Act, as amended (21 U.S.C. 301 et seq.), are the primary authorities for EPA's activities in this area.

Other Federal Agencies with Food Safety Responsibilities

Among the other agencies that play a role in food safety, USDA's Agricultural Research Service (ARS) performs food safety research in support of FSIS's inspection program. It has scientists working in animal disease bio-containment laboratories in Plum Island, NY, and Ames, IA. USDA's Animal and Plant Health Inspection Service (APHIS) indirectly protects the nation's food supply through programs to protect plant and animal resources from domestic and foreign pests and diseases, such as brucellosis and bovine spongiform encephalopathy (BSE, or "mad cow" disease). The Department of Homeland Security (DHS) is to coordinate many food security activities, including at U.S. borders.

Congressional Committees

In the Senate, food safety issues are considered by the Committees on Agriculture, Nutrition, and Forestry; Homeland Security and Governmental Affairs; and Health, Education, Labor, and Pensions. In the House, various food safety activities fall under the jurisdiction of the Committees on Agriculture; Energy and Commerce; Oversight and Government Reform; and Science. Agriculture subcommittees of the House and Senate Appropriations Committees also serve oversight and funding roles in how the major agencies carry out food safety policies.

SELECTED ISSUES

Food safety-related incidents frequently heighten congressional scrutiny of the issue, as a number of developments in 2006 and 2007 illustrated. For example, more than 200 confirmed illnesses and three deaths were linked in fall 2006 to the consumption of bagged fresh spinach grown in California that was found to carry *E. coli* O157:H7. The incident raised public concern about the safety of all fresh leafy produce and stimulated a number of industry and government initiatives to limit future contamination. In February 2007, FDA announced a nationwide recall of peanut butter due to *Salmonella* contamination, after hundreds of illnesses, dating back to August 2006, were linked to the bacterium.[6] Throughout 2007, USDA announced numerous recalls totaling many million pounds of ground beef products due to concerns about *E. coli* O157:H7 contamination.

Attention had shifted to the safety of food imports in early 2007 when adulterated pet food ingredients imported from China sickened or killed numerous dogs and cats and subsequently were found in some hog, chicken, and fish feed.[7] In June 2007, FDA announced that it was detaining imports of certain types of farm-raised seafood from China (specifically, shrimp, catfish, basa, dace, and eel) until their shippers could confirm that they are free of unapproved drug residues.

By late 2007 a series of congressional and Administration reports had been issued that cited significant gaps in the federal food safety system. More than two dozen bills were introduced into the first session of the 110th Congress addressing one or more aspects of the issue. Provisions affecting food safety were included in 2007 in P.L. 110-85, the FDA amendments, including a requirement that FDA establish a registry to which companies must begin to report events involving potentially adulterated foods. Food safety provisions in the 2008 farm bill (P.L. 110-234) include subjecting farmed catfish products to FSIS mandatory inspections similar to those for red meat and poultry; creating an option for state-inspected meat and poultry plants to ship products across state lines; and requiring meat and poultry establishments to notify USDA about potentially adulterated or misbranded products. During the second session, a number of the more comprehensive food safety bills could be marked up. These include bills to impose new requirements, and possibly user fees, on food importers; to establish mandatory recall and/or traceability programs for USDA and FDA-regulated foods; and to increase oversight of domestic food producers, for example. Among the more prominent proposals are separate draft bills being circulated by the chairmen of the House Energy and Commerce Committee and the Senate Health, Education, Labor, and Pensions Committee.

For More Information

For a list of CRS reports providing additional information on food safety issues and legislative proposals, see the Food Safety and Nutrition heading under "Current Legislative Issues: Agriculture," at [http://www.crs.gov/].

End Notes

1 Background on the agencies is updated information that first appeared in CRS Report 98-91, *Food Safety Agencies and Authorities: A Primer* (out of print). Primary sources for that report included various documents and materials provided by federal food safety agencies and by the U.S. Government Accountability Office.

2 Roughly two-thirds of the $1 trillion is for domestically produced farm foods; imports and seafood account for the balance. Source: USDA, Economic Research Service, data accessed January 2008 at the "Food Sector" Web page at [http://www.ers.usda.gov/Browse/FoodSector/].

3 U.S. Department of Health and Human Services, Centers for Disease Control and Prevention, "Foodborne Illness: Frequently Asked Questions," accessed at [http://www.cdc.gov/foodsafety/]. However, this estimate appears to be based primarily on 1997 and earlier data in a report by Paul S. Mead et al., "Food-related Illness and Death in the United States," *Emerging Infectious Diseases*, vol. 5, pp. 607-625, 1999.

[4] *High Risk Series: An Update* (GAO-07-310), January 2007.

[5] Data source: various documents of the Agriculture, Rural Development, Food and Drug Administration, and Related Agencies Subcommittee of the House Committee on Appropriations.

[6] For sources and updates see the FDA website: [http://www.fda.gov/opacom/7alerts.html].

[7] FDA has the same basic safety standards for human foods and animal feeds, including pet food.

In: Food Safety, Fresh Produce and FDA Oversight
Editor: Russell P. Barton
ISBN: 978-1-60692-788-5

Chapter 3

FOOD SAFETY: PROVISIONS IN THE FOOD AND DRUG ADMINISTRATION AMENDMENTS ACT OF 2007

Donna V. Porter

SUMMARY

The Food and Drug Administration Amendments Act of 2007 (FDAAA; P.L.110-85), while primarily concerned with drug and device regulation and their user fees, also contains several provisions on food safety. As enacted, the law requires the Secretary of Health and Human Services to establish, for pet foods, processing and ingredient standards, updated labeling requirements, and an early warning and surveillance system to identify adulteration and associated outbreaks of illness. The Secretary is to work with states to improve the safety of produce and strengthen state food safety programs. The Act requires the Secretary to create a registry for reporting of foods (including human and animal foods) with safety problems, which will help identify the supply chain of the reportable food. Alerts are to be issued for such foods, with records maintained and available for inspection. Additional provisions require attention to aquaculture and seafood inspection, environmental risks associated with genetically engineered seafood products, imported foods, pesticide monitoring, and ginseng dietary supplements.

On September 27, 2007, the President signed the Food and Drug Administration Amendments Act of 2007 (FDAAA; P.L. 110-85, H.R. 3580) into law. The Act primarily addresses issues such as regulation and user fees for prescription drugs and medical devices, clinical trials, pediatric research incentives, and conflicts of interest.[1] However, the Act also contains several provisions on food safety that address recent concerns about contaminated pet food and tracking adulterated foods. This chapter reviews the food safety provisions of the statute, their history, and certain issues surrounding their passage. It will not be updated.

BACKGROUND

The original language of the Senate bill on FDA reform (S. 1082), as introduced, did not contain any provisions on food safety. However, around the time that the Committee on Health, Education, Labor and Pensions (HELP) was considering the bill, health problems with food ingredients from China used in pet food came to the nation's attention. It was ultimately determined that wheat gluten and rice protein concentrate, common pet food ingredients, had been tainted with melamine (an industrial chemical used in the manufacture of plastics, flame-retardants, and other products) to increase thenitrogen content of the products. The contamination of pet food is believed to have led to the death of hundreds of cats and dogs in the United States. Some contaminated feed was also found in feeds given to food-producing animals, but federal officials determined that the situation was unlikely to pose a human food safety risk. Following this news, Senator Durbin and Representative DeLauro introduced the Human and Pet Food Safety Act of 2007 in their respective chambers (S. 1274 and H.R. 2108, respectively) to address the problem of pet and human food contamination. The bills were assigned to committee, but no further action has been taken.

With the FDA legislation on drugs and devices moving forward in the Senate, Senator Durbin introduced as an amendment to S. 1082 several food safety provisions contained in his bill. The amendment passed the Senate 94 to 0 and was sent as part of the FDA reform bill to the House. The FDA reform bill that passed the House, H.R. 2900, did not include any food safety provisions. During the conference on S. 1082 and H.R. 2900, certain changes were made to the legislation, including renumbering it as H.R. 3580 and modifying the final language in the food safety provisions, which are outlined below. The provisions on food safety would primarily establish new statutory authorities. Any amendments to current law are noted, where appropriate.

FOOD SAFETY PROVISIONS IN P.L.110-85

The food safety provisions are contained in Title X of P.L. 110-85, the FDA Amendments Act of 2007. This title generally amends the Federal Food, Drug, and Cosmetic Act (FFDCA), which defines "food" as "articles used for food or drink for man or other animals."[2] As a result, unless otherwise stated in FDAAA, provisions affecting food apply equally to human foods and animal feeds, including pet food.

Section 1001. Findings

This section contains various findings about the safety and integrity of the U.S. food supply, illnesses and deaths caused by contaminated food, the task of preserving the safety of the food supply, and the current level of U.S. food imports and inspections.

Section 1002. Ensuring the Safety of Pet Food

This provision requires that within two years, the Secretary, in consultation with other stakeholders, shall, by regulation, establish processing and ingredient standards for pet food, as well as update nutrition and ingredient labeling on pet food. The Secretary will have one year to establish by regulation an early warning and surveillance system to identify adulteration of the pet food supply and outbreaks of illness associated with pet food. The Secretary shall, in establishing such a system, use surveillance and monitoring mechanisms similar to, or in coordination with, those used by the Centers for Disease Control and Prevention (CDC), consulting with relevant professional organizations and working with existing networks to inform veterinarians and others during a pet food recall.

Section 1003. Ensuring Efficient and Effective Communications During a Recall

During an ongoing recall of human or pet food, the Secretary shall work with relevant stakeholders to collect and aggregate pertinent information through existing networks of communication, including in electronic form, and post information regarding the recall on FDA's website in an easily accessible form.

Section 1004. State and Federal Cooperation

The Secretary shall work with the states on activities and programs that assist in improving the safety of food, including fresh and processed produce, to facilitate coordination and cost-effectiveness. The Secretary shall encourage states to strengthen their food safety programs, especially for retail commercial food establishments, and establish procedures and requirements for ensuring that processed produce is not unsafe for human consumption. The Secretary may provide assistance to states in implementing their food safety programs in the following areas: advisory, technical, training, laboratory, and financial. Under an agreement with a federal, state, or local agency, the Secretary may use, on a reimbursable basis, the personnel, services, and facilities of the agency to assist states.

Section 1005. Reportable Food Registry

The Act amends the FFDCA to create a new section 417 that requires the Secretary, within one year, to establish a Reportable Food Registry within FDA, to which instances of "reportable foods" (excluding infant formula) may be submitted via electronic portal by public health officials and others. Reportable foods are those for which there is a reasonable probability that exposure will cause serious adverse health consequences or death to humans or animals. The Act requires individuals who own or manage FDA-registered food facilities (so-called "responsible parties") to report to FDA within 24 hours if they are aware of an instance of reportable food, and to investigate the cause of the adulteration if the food originated with the responsible party. No report to FDA is required if the responsible party detected the adulteration prior to any transfer of the food or if the adulteration was corrected or the food was destroyed. The Secretary shall promptly review and determine the validity of information submitted for the purpose of identifying reportable food and shall exercise other existing food safety authorities to protect public health. The Secretary shall issue an alert if a food is a reportable food. The Act outlines the data elements required for reporting to FDA for the registry.

Information and records provided to the registry are accessible pursuant to the Freedom of Information Act. Reports to the registry do not constitute admissions that a product caused or contributed to a death, serious injury, or serious illness. Failure to provide a report, or the falsification of a report, is prohibited.

The Secretary shall create a numbering system that allows reports to be linked and amended and that identifies the supply chain for the reportable food. The Secretary shall promptly review a report and require the responsible party to notify appropriate parties in the supply chain and submit a subsequent report or amend a previous report with further information as it becomes available. The Secretary shall share information and coordinate regulatory efforts with the U.S. Department of Agriculture (USDA) for any report on a food product that is within USDA's jurisdiction. The Secretary shall work with state and local public health officials to share information and coordinate regulatory efforts, while reducing duplication, to ensure coverage of the safety of the food supply chain, including food establishments not registered with the FDA (e.g., restaurants). Records must be maintained and available for inspection, if needed. The Secretary of Homeland Security shall be notified immediately if it is suspected that such food may have been deliberately adulterated. Within nine months, the Secretary shall issue guidance to industry about submitting reports to the electronic portal and providing notification to other persons in the supply chain for a reportable food.

Section 1006. Enhanced Aquaculture and Seafood Inspection

The Secretary is authorized to enhance FDA's aquaculture and seafood inspection regime consistent with international agreements and U.S. law. The provision requires a report to Congress within six months of enactment that describes the specifics of the inspection program, the feasibility of developing traceability systems for catfish and seafood products to both foreign and domestic processing plants, and an assessment of the risks associated with contaminants and banned substances. The Secretary may enter into partnerships with the states to implement this inspection program for aquaculture and seafood import products.

Section 1007. Consultation Regarding Genetically Engineered Seafood Products

The FDA Commissioner shall consult with the Assistant Administrator of the National Marine Fisheries Service of the National Oceanic and Atmospheric Administration to produce a report on any environmental risks associated with any genetically engineered seafood products, including the impact on wild fish stocks.

Section 1008. Sense of Congress

This provision states that it is vital for Congress to provide FDA with additional resources, authorities, and direction to ensure the safety of the U.S. food supply, and to provide additional inspectors to improve FDA's ability to safeguard the U.S. food supply. Further, the Secretary should prioritize entering into agreements with U.S. trading partners with regard to food safety, because of the increased volume of international trade in food products, and Congress should work to develop a comprehensive response to the issue of food safety.

Section 1009. Annual Report to Congress

The Secretary shall submit an annual report to the appropriate congressional authorizing and appropriations committees that includes the number and amount of imported food products aggregated by country and type of food, a listing of the number of FDA inspectors of imported food and the number of inspections performed, and aggregate data on the findings of inspections, including data on the violations and enforcement actions used to follow up on the findings and violations.

Section 1010. Publication of Annual Reports

This provision requires the FDA Commissioner to submit to Congress and publish on the FDA website an annual report containing the results of the agency's pesticide monitoring program. The annual report is to include information and analysis similar to that found in the FDA's June 2005 report on pesticide residue monitoring, the results and analysis of the Ginseng Dietary Supplements Special Survey, and certain specified data and other information on interstate and imported shipments of food. Reports are to start with the fiscal years 2004- 2006 combined into a single report. The FDA Commissioner, the Administrator of the USDA Food Safety and Inspection Service (FSIS), the Secretary of Commerce, and the head of the USDA Agricultural Marketing Service (AMS) shall enter into a memorandum of understanding to permit inclusion in their reports of data from FSIS and AMS on meat, poultry, eggs, and certain raw agricultural products.

Section 1011. Rule of Construction

This provision indicates that nothing in the title affects the regulation of, or the adverse reporting system for, dietary supplements created under the Dietary Supplement Health and Education Act of 1994 (P.L. 103-417) or the Dietary Supplement and Nonprescription Drug Consumer Protection Act of 2006 (P.L.109-462), respectively.

Related Activities

The 110th Congress has held numerous hearings on food safety in both chambers because several committees have FDA jurisdiction. While the primary focus has been the safety of imported food, concerns about food safety in general and steps to take in bringing the entire system up-to-date with current science, technology, and globalization of the food supply are the underlying issues. In addition, numerous bills have been introduced in both chambers on various aspects of food safety, several of which have been the subject of the hearings. While all the bills have been assigned to the appropriate committees, none has received any further action, except for the provisions in the FDA Amendments Act of 2007 discussed above.

The Administration has addressed the recent food safety problems in several ways. In May 2007, the FDA Commissioner appointed Dr. David Acheson to the new position of FDA Assistant Commissioner for Food Protection. In this position, he will serve as the coordinator for working with individual FDA product centers and the Office of Regulatory Affairs to coordinate FDA's food safety and security of the food supply assignments and commitments. In addition, Dr. Acheson will serve as the FDA Commissioner's direct liaison to the Department of Health and Human Services and other federal departments and agencies on interagency initiatives on food safety. In November 2007, he completed the integrated strategic plan for protecting the nation's food supply. The Food Protection Plan is organized by the core elements of prevention, intervention, and response to food safety problems, with particular emphasis on prevention. The report describes for each core element the action steps that the agency needs to take and includes legislative proposals needed to fully implement the plan.[3] No time frame for implementation of the plan is provided in the report.

In July 2007, the President appointed a working group on the safety of imported products, including food, to conduct a comprehensive review of current import safety practices and to determine where improvements can be made. This group, headed by HHS Secretary Leavitt, was formed following the revelations about tainted pet food, unsafe tires, antibiotics in seafood, and contaminated toothpaste, all originating in China. Several public meetings and foreign trips were used by the working group to determine the current situation and develop options for the action plan. The group released its Strategic Framework to promote import safety to the President on September 10, 2007.[4] In November 2007, the working group released its Action Plan, which outlined the activities to be completed in 200 days.[5] The action plan contains a number of actions specifically to protect the American food supply, which include the setting of safety standards in the FDA's Food Protection Plan; certification for the seafood inspection program and seafood inspectors stationed in Asian countries; establishment of a security and prosperity partnership for safe food and other products with Mexico and Canada; a memoranda of agreement with China on food and animal feed; foreign training on U.S. safety standards for meat, poultry, and eggs; and a marking requirement on imported food refused entry into the United States to prevent port-shopping. Other actions are planned for non-food imported goods.

ADDITIONAL CRS REPORTS

- RS22600, *The Federal Food Safety System: A Primer*, by Geoffrey S. Becker and Donna V. Porter.
- RL34334, *The Food and Drug Administration: Budget and Statutory History, FY1980-FY2007*, by Judith A. Johnson, Donna V. Porter, Susan Thaul, and Erin D. Williams.
- RL34102, *FDA Legislation in the 110th Congress: A Side-by-Side Comparison of S. 1082 and H.R. 2900*, by Erin D. Williams, Susan Thaul, Sarah A. Lister, Donna V. Porter, and C. Stephen Redhead.
- RL34152, *Food Safety: Selected Issues and Bills in the 110th Congress*, by Geoffrey S. Becker.
- RL34198, *U.S. Food and Agricultural Imports: Safeguards and Selected Issues*, by Geoffrey S. Becker.

END NOTES

[1] Theses issues are described in considerable detail elsewhere. For further information, see CRS Report RL34102, *FDA Legislation in the 110th Congress: A Side-By-Side Comparison of S. 1082 and H.R. 2900*, by Erin D. Williams, Susan Thaul, Sarah A. Lister, Donna V. Porter, and C. Stephen Redhead.

[2] FFDCA § 201(f) [21 U.S.C. § 321(f)].

[3] The Food Protection Plan is available at [http://www.fda.gov/oc/initiatives/advance/food/factsheet.html].

[4] Interagency Working Group on Import Safety, *Protecting Americans Every Step of the Way: A Strategic Framework for Continual Improvement in Import Safety*, report to the President, September 10, 2007, available at [http://www.importsafety.gov/report/index.html].

[5] Interagency Working Group on Import Safety, *The Action Plan for Import Safety: A Road Map for Continued Improvement*, report to the President, November 2007, available at [http://www.importsafety.gov].

In: Food Safety, Fresh Produce and FDA Oversight
Editor: Russell P. Barton
ISBN: 978-1-60692-788-5

Chapter 4

FOOD AND DRUG ADMINISTRATION (FDA): OVERVIEW AND ISSUES

Erin D. Williams

SUMMARY

The Food and Drug Administration (FDA) is the agency within the Department of Health and Human Services (HHS) that regulates human and animal drugs, medical devices, biologics, and most foods. This chapter describes FDA, surveys agency-related issues Congress faces, and cites CRS reports where readers can find more information.

FDA OVERVIEW

FDA is an agency within HHS that regulates a wide range of products valued at more than $1 trillion. (See **Table 1**.) The agency is responsible for the *safety* of most foods (human and animal) and cosmetics, and it regulates both the *safety* and the *effectiveness* of human drugs, biologics (e.g., vaccines, blood and blood components), medical devices, and animal drugs. In many cases, its responsibilities abut those of other agencies. (See **Table 1**.) In such cases, interagency agreements may define the regulatory boundaries.

The primary law authorizing FDA activities is the Federal Food, Drug, and Cosmetic Act (FFDCA; 21 USC Chapter 9). (See **Table 2**.) FDA is also responsible for implementing provisions in other laws, most notably the Public Health Service Act (PHSA; 42 USC Chapter 6A). For example, FDA's authority to regulate most human biologics flows both from the PHSA (§351) and from the FFDCA. (See **Table 2**.)

FDA has three offices that perform agency-wide functions. The Office of the Commissioner conducts overall agency coordination. The Commissioner, FDA's top official, requires Senate confirmation. The Office of Chief Counsel handles the agency's legal needs. FDA's largest office, the Office of Regulatory Affairs (ORA), handles FDA's inspection and enforcement activities. It employs about one-third of the agency's personnel.

FDA's product-specific regulatory responsibilities are handled by five centers: the Center for Biologics Evaluation and Research, the Center for Devices and Radiological Health, the Center for Drug Evaluation and Research, the Center for Food Safety and Applied Nutrition, and the Center for Veterinary Medicine. A sixth center, the National Center for Toxicological Research, conducts scientific research and provides expert technical advice and training that inform FDA's science-based regulatory decisions.

The House and Senate Appropriations subcommittees on agriculture have jurisdiction over FDA's appropriations. FDA's budget consists of two types of funds: public funds appropriated by Congress (called *budget authority* or *direct appropriations*) and private (i.e., industry) funds (called *user fees*).

Table 1. What FDA Does and Does Not Regulate

Product or Activity	Regulatory Agency
Advertising	Federal Trade Commission (FTC) *(FDA regulates prescri-ption drug and restricted device advertising)*
Alcohol	Treasury Department's Bureau of Alcohol, Tobacco, Firearms and Explosives
Biologics	**FDA**
Consumer products *(e.g., toys, cigarette lighters, power tools)*	Consumer Product Safety Commission
Cosmetics	**FDA**
Drinking water	EPA *(**FDA** regulates bottled water)*
Drugs	**FDA** *(Drug Enforcement Administration regulates illegal drug use)*
Foods	**FDA** *(US. Department of Agriculture's (USDA's) Food Safety and Inspection Service regulates most meat and poultry and some egg products)*
Health insurance	Centers for Medicare and Medicaid Services and state authorities
Medical Devices	**FDA**
Organ transplantation	HMS's Organ Procurement Transplantation Network
Pesticides	Environmental Protection Agency (EPA) *(**FDA** and USDA regulate pesticides in food according to EPA's allowable levels)*
Radiation-emitting electronic products	**FDA**
Restaurants and grocery stores	State and local food safety officials
Animal foods, feeds, drugs and devices	**FDA** *(USDA regulates animal biologics)*

Source: Adapted from "What FDA Regulates," at [http://www.fda.gov/comments/regs.html], and "What FDA Does Not Regulate," at [http://www.fda.gov/comments/noregs.html].

Table 2. Location of Subjects within the FFDCA

FFDCA	Subject
Chapter I	Short Title
Chapter II	Definitions
Chapter III	Prohibited Acts and Penalties
Chapter IV	Food
Chapter V	Drugs and Devices
Subchapter A	Drugs and Devices
Subchapter B	Drugs for Rare Diseases and Conditions
Subchapter C	Electronic Product Radiation Control
Subchapter D	Dissemination of Treatment Information
Subchapter E	General Provisions Relating to Drugs and Devices
Subchapter F	New Animal Drugs for Minor Use and Minor Species
Chapter VI	Cosmetics
Chapter VII	General Authority
Subchapter A	General Administrative Provisions
Subchapter B	Colors
Subchapter C	Fees
Subchapter D	Information and Education
Subchapter E	Environmental Impact Review
Subchapter F	National Uniformity for Nonprescription Drugs and Preemption for Labeling or Packaging of Cosmetics
Subchapter G	Safety Reports
Subchapter H	Serious Adverse Event Reports
Subchapter I	Reagan-Udall Foundation for the Food and Drug Administration
Chapter VIII	Imports and Exports
Chapter IX	Miscellaneous

FDA-Related Issues

FDA-related issues of interest to Congress generally rest on the central question of how best to give people access to useful products while protecting them from unsafe ones. Creating too many regulatory requirements raises costs and prevents products from reaching consumers. Creating too few places consumers at risk.[1]

The 110th Congress placed a new focus on FDA's regulatory responsibilities, passing the most comprehensive FDA reform legislation in almost a decade: the Food and Drug Administration Amendments Act of 2007 (FDAAA; PL 110-85).[2] FDAAA reauthorized four expiring programs and expanded the agency's authority to regulate the safety of prescription drugs and biologics, medical devices, and foods. In the wake of that legislation, issues remain both in areas that FDAAA did not comprehensively address and in areas raised by its implementation. The following is an introduction to the types of issues that Congress now faces with respect to FDA. For further assistance with any FDA-related issue, see **Table 3** (at the end of the report) for a list of CRS experts.

Budget

The primary budget-related question faced by Congress is how to fund the agency sufficiently for it to carry out its responsibilities, while also funding competing national needs, and ensuring that the agency operates cost-effectively.[3] Some secondary budget-related questions center on user fees.[4] They ask to what extent FDA should be funded by money from the industries it regulates, and for which activities such funds should be collected and used (e.g., premarket review, inspection and enforcement).

Premarket Approval

Before FDA will permit drugs, devices, and biological products to be marketed in the United States, the agency requires evidence that they are safe and effective. (Only limited types of food ingredients require premarket approval.) Premarket approval processes vary by product type.[5] Most processes rely on evidence from clinical trials. The topic of clinical trials raises questions of when it is appropriate to test new products on people, particularly on children, and in what circumstances it is appropriate to publicize the trials and their results.[6]

The approval process for new products can take time. While this may be of little consequence for people with manageable conditions, special issues arise for people with life-threatening diseases or conditions for which there is no current treatment. As a result, some interest has been focused on mechanisms for giving people access to unapproved medications, and for speeding FDA's approval process.[7]

Products on the Market

FDA is responsible for ensuring the safety of products it regulates — including foods — once they are on the market. It accomplishes this goal through product tracking, inspection, and enforcement. Some attention has been focused on the fact that the agency has different enforcement authorities for product types. For example, FDA has mandatory recall authority for medical devices and infant formula, but not for other foods or for prescription drugs.[8] Questions have also arisen regarding whether marketing with FDA approval should preempt certain tort claims.[9]

FDA's role dovetails with product safety issues that cut across numerous agencies, creating the need for interagency coordination.[10] For products such as tobacco and genetic tests, the current patchwork of regulation — or lack thereof — has led to calls for comprehensive FDA oversight.[11] In areas of shared responsibility, such as product importation and advertising, FDA's role, and its ability or willingness to use agency resources to fulfil its responsibilities, has caused concern.[12]

Food Safety

As noted above, there is no premarket approval for foods or most food ingredients. FDA's statutory authority and historical approach are reactive, focused on foods or ingredients that are found to be unsafe.[13] Many policy makers seek a more preventive approach and debate how to craft such a system. Proposals include having FDA inspect processes instead of products, set performance measures, or increase industry's burden to assure safety. A successful approach may take into account the variety of foods FDA regulates, a growing stream of imported foods, limited global food tracking systems, and the agency's finite resources.

Advisory Committees

In its vetting of the numerous products it regulates, FDA relies on non-binding input from groups of outside experts known as *advisory committees*. Because the experts in specialized fields may often be those with a financial stake in the resulting products, questions have emerged about managing conflicts of interest in the advisory committees.[14]

Products and Technologies

Questions have been raised about FDA's ability to keep up with the increasing sophistication of some types of products it regulates, such as genetic tests, follow-on (generic) biologics, and cell- and tissue-based products.[15] A similar concern has been raised about its ability to assess health threats that may arise from combined exposures to multiple types of FDA-regulated products, and other exposures.[16] Others have focused on politically sensitive products, such as the contraceptive "Plan B."[17] There are also questions about the adequacy of FDA's assessment of the safety of products produced using emerging technologies, such as biotechnology.[18] All of the above questions are intensified for *combination products* — those composed of two or more regulated components (e.g., a drug/device, or a biologic/device) — whose regulation requires coordination across FDA centers.

Table 3. CRS Experts

Area	Analyst(s) and Phone Number(s)
FDA Team Leader	Erin D. Williams (7-4897)
Foods	Donna V. Porter (7-7032), Geoffrey S. Becker (7-7287)
Human Drugs	Susan Thaul (7-0562)
Biologics	Judith A. Johnson (7-7077)
Animal Drugs and Feeds	Sarah A. Lister (7-7320)
Devices and Radiological Health	Erin D. Williams (7-4897)
Blood and Plasma Products	C. Stephen Redhead (7-2261)
Human Cellular and Tissue Products	Bernice Reyes-Akinbileje (7-2260), Erin D. Williams (7-4897)
Legal Issues	Vanessa K. Burrows (7-0831)

END NOTES

[1] See CRS Report RL33802, *Pharmaceutical Costs: A Comparison of Department of Veterans Affairs (VA), Medicaid, and Medicare Policies,* by Gretchen A. Jacobson, Sidath Viranga Panangala, and Jean Hearne, and

"Drugs, Biologics, and Medical Devices," *CRS CLI,* at [http://apps.crs.gov/cli/cli.aspx?PRDS_CLI ITEM ID=2678&from=3&fromId=13].

[2] See CRS Report RL34465, *FDA Amendments Act of 2007 (P.L. 110-85)*, by Erin D. Williams and Susan Thaul, and CRS Report RS22779, *Food Safety: Provisions in the Food and Drug Administration Amendments Act of 2007*, by Donna V. Porter.

[3] See CRS Report RL34334, *The Food and Drug Administration: Budget and Statutory History, FY1980-FY2007*, by Judith A. Johnson, Donna V. Porter, Susan Thaul, and Erin D. Williams, and CRS Report RL34638, *The FDA FY2009 Budget*, by Judith A. Johnson, Sarah A. Lister, Donna V. Porter, Pamela W. Smith, Susan Thaul, and Erin D. Williams.

[4] See CRS Report RL33914, *The Prescription Drug User Fee Act (PDUFA): History, Reauthorization in 2007, and Effect on FDA*, by Susan Thaul (hereinafter RL33914); CRS Report RL34571, *Medical Device User Fees and User Fee Acts*, by Erin D. Williams; and CRS Report RL34459, *Animal Drug User Fee Programs*, by Sarah A. Lister (hereinafter RL34459).

[5] See RL33914; RL34459; and CRS Report RL32826, *The Medical Device Approval Process and Related Legislative Issues*, by Erin D. Williams (hereinafter RL32826).

[6] See CRS Report RL32909, *Federal Protection for Human Research Subjects: An Analysis of the Common Rule and Its Interactions with FDA Regulations and the HIPAA Privacy Rule*, by Erin D. Williams; CRS Report RL33986, *FDA's Authority to Ensure That Drugs Prescribed to Children Are Safe and Effective*, by Susan Thaul; and CRS Report RL32832, *Clinical Trials Reporting and Publication*, by Erin D. Williams.

[7] See CRS Report RS22814, *FDA Fast Track and Priority Review Programs*, by Susan Thaul.

[8] See CRS Report RL34167, *The FDA's Authority to Recall Products*, by Vanessa K. Burrows.

[9] See *Riegel v. Medtronic, Inc.* (552 U.S. __ (2008); No. 06-179 (U.S. February 20, 2008)).

[10] See "Product Safety Authorities and Remedies," *CRS CLI*, at [http://apps.crs.gov/cli/cli.aspx?PRDS_CLI_ITEM_ID=3117&from=3&fromId=13].

[11] See CRS Report RL32619, *FDA Regulation of Tobacco Products: A Historical, Policy, and Legal Analysis*, by C. Stephen Redhead and Vanessa K. Burrows; CRS Report RL33719, *Tobacco: Selected Legal Issues*, by Vanessa K. Burrows; CRS Report RS22944, *Federal Trade Commission Guidance Regarding Tar and Nicotine Yields in Cigarettes*, by Vanessa K. Burrows; and CRS Report RL33832, *Genetic Testing: Scientific Background for Policymakers*, by Amanda K. Sarata (hereinafter RL33832).

[12] See CRS Report RL32191, *Prescription Drug Importation and Internet Sales: A Legal Overview*, by Vanessa K. Burrows; CRS Report RS21711, *Legal Issues Related to Prescription Drug Sales on the Internet*, by Vanessa K. Burrows; and RL32826.

[13] See CRS Report RS22600, *The Federal Food Safety System: A Primer*, by Geoffrey S. Becker and Donna V. Porter, and "Food Safety and Nutrition," *CRS CLI*, at [http://apps.crs.gov/cli/level_2.aspx?PRDS_CLI_ITEM_ID=13].

[14] See CRS Report RS22691, *FDA Advisory Committee Conflict of Interest*, by Erin D. Williams.

[15] See RL33832; CRS Report RL34045, *FDA Regulation of Follow-On Biologics*, by Judith A. Johnson; CRS Report RL33901, *Follow-On Biologics: Intellectual Property and Innovation Issues*, by Wendy H. Schacht and John R. Thomas; CRS Report RL34614, *Nanotechnology and Environmental, Health, and Safety: Issues for Consideration*, by John F. Sargent; CRS Report RL34332, *Engineered Nanoscale Materials and Derivative Products: Regulatory Challenges*, by Linda-Jo Schierow; CRS Report RL33540, *Stem Cell Research: Federal Research Funding and Oversight*, by Judith A. Johnson and Erin D. Williams; and CRS Report RL33554, *Stem Cell Research: Ethical Issues*, by Erin D. Williams and Judith A. Johnson.

[16] See CRS Report RL34572, *Phthalates in Plastics and Possible Human Health Effects*, by Linda-Jo Schierow and Margaret Mikyung Lee, and CRS Report RS22869, *Bisphenol A (BPA) in Plastics and Possible Human Health Effects*, by Linda-Jo Schierow and Sarah A. Lister.

[17] See CRS Report RL33728, *Emergency Contraception: Plan B*, by Judith A. Johnson and Vanessa K. Burrows.

[18] CRS Report RL33334, *Biotechnology in Animal Agriculture: Status and Current Issues*, by Geoffrey S. Becker and Tadlock Cowan.

In: Food Safety, Fresh Produce and FDA Oversight
Editor: Russell P. Barton

ISBN: 978-1-60692-788-5

Chapter 5

STATEMENT OF DAVID ACHESON, BEFORE THE COMMITTEE ON AGRICULTURE SUBCOMMITTEE ON HORTICULTURE AND ORGANIC AGRICULTURE UNITED STATES HOUSE OF REPRESENTATIVES

Department of Health & Human Services

INTRODUCTION

Good afternoon, Chairman Cardoza and Members of the Subcommittee. I am Dr. David Acheson, Associate Commissioner for Foods at the Food and Drug Administration (FDA or the Agency), which is part of the Department of Health and Human Services (HHS). I am pleased to be here today with my colleague, Dr. Lonnie J. King, from the Centers for Disease Control and Prevention (CDC), which is also part of HHS. FDA appreciates the opportunity to discuss the recent foodborne illness outbreak associated with fresh produce contaminated with *Salmonella* Saintpaul and the measures we are taking to enhance the safety of fresh produce and to enhance traceability.

FDA is the Federal agency that regulates almost everything we eat except for meat, poultry, and processed egg products, which are regulated by our partners at the United States Department of Agriculture (USDA). FDA is committed to ensuring that America's food supply continues to be among the safest in the world.

There is no question that the *Salmonella* Saintpaul outbreak investigation has been one of the most complex investigations in recent memory. I assure you that FDA is committed to working with all our food safety partners to examine ways to remove or mitigate some of the complicating factors to expedite tracebacks. In my testimony, I will discuss some of the factors that made this investigation so complex. I will also describe some of the challenges we face both in preventing fresh produce from becoming contaminated in the first place and in investigating outbreaks associated with fresh produce. I will also discuss some of the specific measures FDA is taking to enhance the safety of fresh produce and other foods to prevent future outbreaks and to improve traceability when an outbreak occurs.

Food can become contaminated at many different steps – on the farm, in processing or distribution facilities, during transit, at retail and food service establishments, and in the home. In recent years, we have done a great deal to prevent both intentional and unintentional contamination of food at each of these steps. FDA has worked with other Federal, state, local, tribal, and foreign counterpart food safety agencies, as well as with law enforcement and intelligence-gathering agencies, and with industry, consumer groups, and academia to significantly strengthen the nation's food safety and food defense system across the entire distribution chain.

This cooperation has resulted in greater awareness of potential vulnerabilities, the creation of more effective prevention programs, new surveillance systems, and the ability to respond more quickly to outbreaks of foodborne illness. However, changes in consumer preferences, changes in industry practices, and the rising volume of imports posed challenges that required us to adapt our current food protection strategies and to develop the Food Protection Plan and the Action Plan for Import Safety, which I will discuss later in my testimony.

CHALLENGES OF FRESH PRODUCE

The number of illnesses associated with fresh produce is a continuing concern for FDA, and we have worked on a number of initiatives to reduce the presence of pathogens in these foods.

Fresh produce presents special challenges. For example, consumption of produce, particularly "ready-to-eat" products, has increased dramatically during the past decade. This is a positive development from a nutrition perspective, but also a new dynamic that challenges our food safety efforts.

Because most produce is grown in an outdoor environment, it is vulnerable to contamination from pathogens that may be present in the soil, in agricultural or processing water, in manure used as fertilizer, or due to the presence of animals in or near fields or packing areas. Produce also may be vulnerable to contamination due to inadequate worker health and hygiene protections, environmental conditions, inadequate production safeguards, and inadequate sanitation of equipment and facilities. Fresh produce is produced on tens of thousands of farms, and contamination at one step in the growing, packing, and processing chain can be amplified throughout the subsequent steps. The fact that produce is often consumed raw or with only minimal processing, without any type of intervention that would eliminate pathogens (if they are present) prior to consumption, contributes to its potential as a source of foodborne illness.

Consequently, addressing the way fresh produce is grown, harvested, and moved from field to fork is crucial to minimizing the risk of microbial contamination. In recent years, FDA has initiated several activities to address safety concerns associated with the production of fresh produce. Some of these activities include: working with industry to develop guidance on ways to prevent or minimize potential contamination, conducting educational outreach to consumers on safe food handling practices, sampling and analyzing both domestic and imported produce for pathogens, and working with industry and foreign countries to promote the use of good growing, harvesting, packing, transporting, and processing practices. For example, just last month, FDA provided training in good agricultural practices in Costa Rica.

Research is also a critical element of our efforts to improve the safety of fresh produce. Our current research agenda is focused on improving the identification and detection of disease- causing bacteria and toxins in a variety of foods. More rapid and precise testing methods to identify contaminants are important for detecting contamination if it is present and minimizing the spread of foodborne disease once it occurs. In addition, we are working with academia, industry, other Federal agencies, and state governments to develop both risk-based microbiological research programs and technology transfer programs to ensure that the latest food technology reaches the appropriate end users along the supply chain.

I would now like to provide a brief description of the typical traceback process.

Traceback Process

Once CDC, through its epidemiological investigation which involves working with state and local governments, identifies the possible food(s) associated with a foodborne illness outbreak, CDC notifies FDA. At that point, we start our traceback investigation to identify the source of the contamination. We work with industry and with local, state, and Federal officials, and, when needed, with foreign governments, to identify the source of the contamination. We do this by tracing the food suspected of being the vehicle for transmitting the pathogen back through the supply chain from the retailer or restaurant and inspecting or investigating points throughout the supply chain to determine where the contamination most likely occurred. Tracing food requires us to find and examine documentation (such as bills of lading and invoices) for the product throughout the supply chain. We also obtain information on the practices and conditions under which the product was stored and handled at each point to better determine shipments of interest and whether contamination may have occurred at each point.

Traceback investigations involving fresh produce are more difficult because the food is perishable and is usually no longer available for testing by the time consumers become ill. In addition, fresh fruits and vegetables are often sold loose without any packaging that could provide information about its source. Further, practices such as packing or repacking produce from multiple sources add complexity to traceback investigations. As each traceback investigation is different, I would like to mention three recent examples which illustrate the different degrees of difficulty.

Peanut Butter

In 2007, CDC notified FDA of a multi-state outbreak of *Salmonella* Tennessee infections associated with the consumption of peanut butter. In this case, because it was not a perishable food, consumers who had become ill still had jars of peanut butter available for testing. This enabled investigators to confirm the presence in that food of the contaminant associated with the outbreak. Further, because the food was packaged, the investigators were able to identify the manufacturer through the information on the jars. This is an example of a rapid traceback in which the necessary information was readily available.

Fresh Spinach

In 2006, CDC informed FDA of a multi-state outbreak of illnesses associated with the consumption of fresh spinach contaminated with *Escherichia coli* O157:H7. Although this outbreak involved a perishable food, the food was sold in a package. The traceback investigation was facilitated because several consumers who had become ill still had packages of fresh spinach in their refrigerators. The information on those packages ultimately led investigators to the spinach processors. By looking at the processor's records, the investigators were able to identify the implicated farms associated with the identified production lot of bagged spinach. This is an example of a traceback of medium complexity that took a little longer than the peanut butter example but which was aided by the information on the package.

Salmonella Saintpaul

The current outbreak investigation, which initially focused on certain types of raw tomatoes, provides an example of one of the most difficult traceback investigations. On May 26, CDC informed FDA of the hypothesis of a possible association between ill persons and the consumption of raw tomatoes. On May 31, CDC formally notified FDA of a significant statistical association between consumption of certain types of tomatoes and a multi-state outbreak of *Salmonella* Saintpaul infections, and FDA decided to initiate investigations attempting to trace the tomatoes reported to have been eaten by ill persons back to their sources. Raw tomatoes are a perishable commodity and, thus, are unlikely to be in the consumer's home after the consumer becomes ill, obtains a diagnosis, and a foodborne illness outbreak is identified. Further, raw tomatoes are often sold loose, without any form of packaging. In this case, we learned that many tomatoes had been shipped to washing, packing, and repacking facilities where they were or might have been commingled with other tomatoes from many different sources. This commingling has the potential to multiply the quantity of food that is contaminated. It also increases the difficulty in determining which tomatoes were the source of the illnesses. A further complicating factor was caused by entities in the supply chain using different terminology to describe the tomatoes. For example, one party might describe the tomatoes as "hothouse" or "greenhouse" tomatoes while the next party in the chain might describe them simply as "tomato bulk." Yet another party might use a descriptor such as "green six-by-six." This lack of consistency in nomenclature makes it more difficult and more time-consuming to connect the links in the chain and to identify the source of the tomatoes.

SALMONELLA SAINTPAUL OUTBREAK INVESTIGATION

Since May 31, many FDA employees in the field and at headquarters have been working continuously on the outbreak investigation to identify the source(s) of the illnesses. To help the public distinguish tomatoes not associated with the outbreak, FDA adopted the policy of specifically designating the types of tomatoes implicated in the outbreak as well as listing growing areas that were not part of the outbreak. Based on information provided by CDC, state officials, and from our own investigations, FDA has been regularly updating the information on its website, conducting media calls, and updating our Federal, state, and local partners, along with the affected industries.

As is our usual course, FDA's recommendations for consumers were focused on protecting public health and were based on epidemiological information from the state agencies and CDC.

From them we learned initially that illness was statistically linked to consumption of raw tomatoes. Ill persons reported consuming red round, red plum, and red Roma tomatoes. Because few ill persons had reported consuming other types of tomatoes, we advised consumers that these other types of tomatoes had not been implicated. We also had information from our ongoing traceback investigation that a limited number of geographic regions were being identified as possible sources of the tomatoes that were associated with the outbreak. A number of states informed FDA that growers within their jurisdictions either were not shipping tomatoes during the period of concern or they would not have shipped tomatoes as widely as would have been required to account for this multi-state outbreak. This aggregated information allowed us to advise consumers that they could eat certain types of tomatoes and all tomatoes from a number of countries and states (or from certain regions within a state) with confidence that they were not from the sources that were identified in the traceback investigation.

On June 30, CDC advised FDA that their epidemiological data from the ongoing outbreak indicated that jalapeño and Serrano peppers also might be implicated in the outbreak. Accordingly, on July 1, FDA expanded its investigation into peppers as well and advised consumers at increased risk of complications from infection (elderly persons, infants, and persons with impaired immune systems) not to consume raw Serrano and jalapeño peppers.

On July 17, FDA lifted its warning to consumers to avoid certain types of raw tomatoes. FDA announced that tomatoes currently on the market are not considered to be a possible source of the continuing *Salmonella* Saintpaul illnesses because the tomatoes coming to market now are harvested from different growing areas than those initially implicated. We also reiterated our recommendation to consumers at increased risk of infection to avoid eating Serrano and jalapeño peppers while the investigation continues.

On July 21, FDA announced that one of the jalapeño pepper samples we tested is a genetic match with the outbreak serotype, *Salmonella* Saintpaul. This finding is strong evidence that jalapeño peppers were involved in the outbreak; however, it does not exonerate other foods. While this one positive sample does not provide the whole story, this genetic match is an important break in the case that we hope will help us pinpoint the source of the contamination. FDA obtained the jalapeño pepper sample during an inspection of the Agricola Zaragoza produce distribution center in McAllen, Texas. The company voluntarily issued a recall. The pepper was grown in Mexico, but that did not mean the pepper was contaminated in Mexico. We continued to investigate the source of the contamination.

Based on this finding, on July 21, FDA advised consumers to avoid eating fresh jalapeño peppers and foods made with them. This advisory did not include cooked or pickled jalapeño peppers. As the traceback investigation continued into the source of the pepper's contamination, the review of the current traceback investigation and harvesting dates, matched with the dates that people became ill, combined to indicate that the contaminated jalapeño pepper originated in Mexico and not at the plant in Texas. Therefore, on July 25, FDA announced that there was no indication that domestically grown jalapeño or Serrano peppers are implicated in the outbreak. We updated our consumer advisory to indicate that our advice to avoid raw jalapeño and Serrano peppers now applies only to peppers grown, harvested, or packed in Mexico. In addition to domestically grown raw jalapeño and Serrano peppers, canned, pickled, and cooked jalapeño and Serrano peppers from any and all geographic locations also are not connected with this outbreak. Serrano and jalapeño peppers are often grown together, are often served in the same foods, and often travel along the same distribution routes. The finding of the contaminated jalapeño pepper does not mean that Serrano peppers were not also associated with the outbreak.

We are working with state regulatory agencies and the food industry, including restaurants, grocery store chains, and wholesalers to ensure that this new, more narrowly focused advisory is clearly understood by everyone. Our investigation into the source of the contamination is ongoing. We will continue to refine our consumer guidance as our investigation continues.

I would now like to describe some of our recent activities to improve traceability of fresh produce.

Recent FDA Activities to Improve Traceability of Fresh Produce

The ability to trace pathways of any food, including tomatoes and other fresh produce, through every point in the supply chain is crucial for limiting foodborne illness in an outbreak, for preventing future outbreaks, and for reducing the impact on the segments of the industry whose products were not associated with the illnesses. The pathways that fresh produce travels from field to consumer have become increasingly complex, with items sometimes changing hands many times in the supply chain.

FDA formed an internal multi-Center group to meet with external entities (such as industry, consumers, and Federal, state, local, and foreign governments) to better understand the universe of track and trace systems that are currently in use or being developed. FDA has reached out to various organizations, including trade associations and consumer groups, to gain a better understanding of best industry practices for traceability, including the use of electronic and other technologies that speed and enhance the traceback process and the use of systems that connect all the links in the produce supply chain. FDA is using this information to develop recommendations for the fresh produce industry to use to improve its internal traceback systems. We plan to hold a public meeting in the fall to further the exchange of information on available technology and best practices for enhanced traceability.

We have been working extensively with states and the fresh produce industry to encourage incorporation of traceability procedures and technology. For example, FDA assisted the Florida Tomato Commission and the University of Florida/Institute of Food and Agricultural Sciences in the development of Florida's Tomato Best Practices Manual. This Manual incorporates Good Agricultural Practices, Good Handling Practices, and traceability recommendations for industry. The Manual formed the basis of the State of Florida's tomato safety rule.

Another recent example is the final guidance for the fresh-cut produce industry, which FDA issued this year. The guidance includes a section on tracebacks and a section on documentation and recordkeeping. FDA also has provided industry its "Guide to Traceback of Fresh Fruits and Vegetables Implicated in Epidemiological Investigations," which is used by our investigators.

Last month, FDA issued a Request for Applications to provide funding to six states to establish Food Protection Rapid Response Teams to investigate multi-state outbreaks of foodborne illness. Enhancing the infrastructure of state food protection programs and strengthening joint Federal/state responsiveness at a local level are an important way to protect consumers by expediting traceback investigations.

We will continue to work with Federal, state, local and international food safety partners and with industry to develop guidance, conduct research, develop educational outreach materials, and initiate other commodity-, practice-, or region-specific programs to enhance the safety of fresh produce.

ACTION PLAN FOR IMPORT SAFETY AND FOOD PROTECTION PLAN

To enhance safety across the range of imported consumer products, last November, Secretary Leavitt presented to the President an Action Plan for Import Safety (Action Plan) which reflects the input of twelve Departments and Agencies and provides recommendations to enhance the safety of imported products. In conjunction with the Action Plan, FDA released the Food Protection Plan, which provides a framework to identify and counter potential hazards with respect to both domestic and imported food. Achieving the food safety enhancements identified by these plans will require the involvement of all our food safety partners – Federal, state, local, tribal, and foreign governments; industry; academia; consumers; and Congress. Both Plans build in safety measures across a product's life cycle, from the time a food is produced to the time it is distributed and consumed. They encompass three core elements: prevention, intervention, and response.

The Food Protection Plan identified ten legislative authorities necessary for achieving full implementation. We encourage Congress to provide these authorities, which would:

- Allow FDA to require preventive controls against intentional adulteration at points of high vulnerability in the food chain;
- Authorize FDA to issue additional preventive controls for certain high-risk foods;
- Require food facilities to renew their FDA registrations at least every two years and allow FDA to modify the current food product categories for purposes of registration;
- Authorize FDA to accredit highly-qualified third parties for voluntary food inspections;
- Require a new reinspection fee from facilities that fail to meet current Good Manufacturing Practice (cGMPS) requirements;
- Empower FDA to require electronic import certificates for shipments of designated high-risk products from countries with which FDA has concluded an agreement on a

certification program that provides a level of safety sufficient to meet FDA standards;

- Allow FDA to charge export certification fees for food and animal feed to improve the ability of U.S. firms to export their products;
- Authorize FDA to refuse admission of imported food if FDA inspection access is delayed, limited or denied;
- Empower FDA to issue a mandatory recall of food products if voluntary recalls are not effective; and
- Give FDA enhanced access to food records during emergencies.

Last month, the Secretary announced that the Administration is increasing its Fiscal Year (FY) 2009 budget request for FDA by $275 million. This increase brings the Administration's total proposed increase in FDA's budget, including user fees, for FY 2009 to $406.3 million, a 17.9% increase over FY 2008. A large portion of this increase ($125 million) will be used for food safety and will allow FDA to intensify actions to implement the Food Protection Plan. This is in addition to the $42.2 million increase proposed for food protection in the budget announced in February 2008.

On June 30, the President signed the FY 2008 Supplemental Appropriation into law. This appropriation act provided $150 million for FDA, and these resources will allow FDA to accelerate its transformation of its regulatory strategies to meet the challenges of the evolving global marketplace for food and medical products. The funds in the supplemental appropriations act will allow FDA to further implement the Food Protection Plan, the Action Plan for Import Safety, and important medical product priorities. It will specifically allow FDA to expand its food safety activities, such as increasing inspections, performing research on mechanisms of food contamination, establishing offices overseas to build capacity with our foreign partners, developing and validating more rapid detection tools, enhancing our information technology systems to support interoperable databases, and enhancing FDA's ability to identify and target the greatest threats from intentional and unintentional contamination.

CONCLUSION

FDA is working hard to ensure the safety of food, in collaboration with its Federal, state, local, tribal, and international food safety partners, and with industry, consumers, and academia. As a result of this effective collaboration, the American food supply continues to be among the safest in the world. However, the *Salmonella* Saintpaul foodborne illness outbreak underscores the challenges we face. Once our investigation has determined the cause of the *Salmonella* contamination, we will examine what other measures are needed.

In the meantime, we have been making progress and are moving forward to implement the Plans. We recently issued 6-month updates that demonstrate the specific actions we have been taking to implement the Plans. For example, we have formed a Risk-Based Steering Committee with the charge of ensuring that a comprehensive risk-based approach is taken with regard to food protection. We are holding a 50-state meeting in August to share information and develop strategies for implementing the Food Protection Plan and to enhance future collaborations between Federal, state, and local partners. Progress also has been made in identifying food vulnerabilities and mitigation strategies; for example, FDA has identified several natural plant bacteria that are effective in preventing contamination of tomatoes with *Salmonella* Newport. FDA scientists received training and instruments to rapidly detect and accurately identify *Salmonella* serovars using a new molecular method. We have strengthened the response to food safety threats by providing incident command system training to our FDA offices around the country and to states and by developing templates to enhance communication during a food recall. We will continue to strive to reduce the incidence of foodborne illness to the lowest level possible.

Thank you for the opportunity to discuss FDA's continuing efforts to enhance food safety and traceability. I would be happy to answer any questions.

In: Food Safety, Fresh Produce and FDA Oversight
Editor: Russell P. Barton

ISBN: 978-1-60692-788-5

Chapter 6

STATEMENT OF DAVID ACHESON, BEFORE THE COMMITTEE ON AGRICULTURE SUBCOMMITTEE ON HORTICULTURE AND ORGANIC AGRICULTURE UNITED STATES HOUSE OF REPRESENTATIVES

Department of Health & Human Services

INTRODUCTION

Good morning, Chairman Cardoza and Members of the Subcommittee. I am Dr. David Acheson, Assistant Commissioner for Food Protection at the Food and Drug Administration (FDA or the Agency), which is part of the Department of Health and Human Services (HITS). I am pleased to be here today with my colleague, Mr. Lloyd Day of the U.S. Department of Agriculture (USDA). FDA appreciates the opportunity to discuss the recent outbreaks of foodborne illness associated with fresh produce and the measures we are taking to enhance the safety of these products.

FDA is committed to ensuring that America's food supply continues to be among the safest in the world. In recent years, we have done a great deal to protect the food supply from both unintentional and deliberate contamination. We have made significant progress in both, but the recent outbreaks of foodborne illness underscore the need to renew our focus on multidisciplinary and integrated food safety strategies.

The Commissioner of Food and Drugs, Dr. Andrew von Eschenbach, recently appointed me to the newly created position of Assistant Commissioner for Food Protection. My first priority in this position is to develop a new strategy for food safety and food defense that will address changes in the global food safety and food defense system, identify our most critical needs, and serve as a framework to help us address the challenges we face. The goal is to ensure a comprehensive and robust food safety and food defense program that is tailored to meet the risks posed by the types of foods we regulate and that focuses on prevention, ensures compliance with preventive controls, and rapidly responds when contaminated food or feed is detected, or when there is harm to humans or animals.

The number of illnesses associated with fresh produce is a continuing concern of the Agency, and we have worked on a number of initiatives to reduce the presence of pathogens in these foods. Fresh produce presents special challenges. For example, consumption of produce, particularly "ready-to-eat" products, has increased dramatically during the past decade. This is a positive development from a nutrition perspective, but also a new dynamic that challenges our food safety efforts.

Because most produce is grown in an outdoor environment, it is vulnerable to contamination from pathogens that may be present in the soil, in agricultural or processing water, and in manure used as fertilizer, or due to the presence of animals in or near fields or packing areas. Produce also is vulnerable to contamination due to inadequate worker health and hygiene protections, environmental conditions, inadequate production safeguards, and inadequate sanitation of equipment and facilities. The fact that produce is often consumed raw or with only minimal processing, without any type of intervention that would reduce or eliminate pathogens prior to consumption, contributes to its potential as a source of foodborne illness. Consequently, addressing the way fresh produce is grown, harvested, and moved from field to fork is crucial to minimizing the risk of microbial contamination.

In my testimony today, I will describe HHS's role concerning activities related to food safety. Then, I will describe some of the specific efforts that FDA is taking to enhance the safety of fresh produce to prevent future outbreaks.

HHS's Role in Food Safety

FDA's primary mission is to protect the public health. Ensuring that FDA-regulated products are safe and secure is a vital part of that mission. FDA is the Federal agency that regulates everything we eat except for meat, poultry, and processed egg products, which are regulated by our partners at USDA.

Although FDA has the lead responsibility within HHS for ensuring the safety of food products, the Centers for Disease Control and Prevention (CDC) has an important complementary and non-regulatory public health role. CDC is the lead Federal agency for conducting disease surveillance and outbreak investigation and routinely monitors the occurrence of specific illnesses in the U.S. attributable to the entire food supply. The disease surveillance systems coordinated by CDC, in collaboration with states, provide an essential early-information network to detect dangers in the food supply and to reduce foodborne illness. Two key surveillance components of our nation's early information network are PulseNet and OutbreakNet. PulseNet is a national network of public health laboratories that perform DNA fingerprinting on foodborne bacteria that result in human illness. The PulseNet network permits rapid comparison of these fingerprint patterns through an electronic database at CDC. OutbreakNet is a network of public health epidemiologists who, under CDC's coordination, investigate suspected foodborne disease outbreaks to determine which foods may be involved and, thus, which control strategies may be needed. CDC's ability to detect and investigate outbreaks of foodborne illness through its networks enables CDC to alert FDA and USDA about implicated food products associated with foodborne illness. CDC also provides expert scientific evaluations of the effectiveness of foodborne disease prevention strategies.

FDA contributes financially and scientifically to the Foodborne Diseases Active Surveillance Network (FoodNet), the principal foodborne disease component of CDC's Emerging Infections Program (EIP). FoodNet is a collaborative activity of CDC, FDA, the Food Safety and Inspection Service (FSIS) of USDA, and ten EIP sites. Through this active surveillance system, these sites actively seek out information on foodborne illnesses identified by clinical laboratories, collect information from patients about their illnesses, and conduct investigations to determine which foods are linked to specific pathogens. This surveillance system provides important information about changes over time in the burden of foodborne diseases. For example, data from FoodNet help public health and food safety agencies evaluate the effectiveness of current food safety initiatives and plan future food safety activities to prevent and reduce emerging foodborne illnesses.

In addition to working closely with CDC, our sister public health agency, FDA has many other food safety partners – Federal, state, and local agencies; international food safety partners; academia; and industry.

INITIATIVES TO ENHANCE PRODUCE SAFETY

To reduce the risk of foodborne illness at all points in the food chain, FDA has adopted a "farm-to-fork" approach to food safety. This approach systematically applies risk management principles at each step as food moves from growers and producers to consumers. While FDA has been working to enhance produce safety for a number of years, the Agency has sharpened its focus in response to the recent produce-related outbreaks.

FDA has focused its food safety efforts in three key areas:

- strengthening the scientific basis for FDA's food safety program with an emphasis on prevention;
- enhancing effective partnerships; and
- improving risk-based targeting of inspection resources.

I will elaborate on these below.

Strengthening the Scientific Basis for FDA's Program to Improve Food Safety

Strengthening the scientific basis for FDA's program to improve food safety is essential to improving FDA's effectiveness at protecting public health. FDA's existing research program includes activities involving microbiological contamination, chemical contaminants, biotechnology/allergenicity issues, seafood safety, dietary supplements safety, color additives safety, and consumer studies. The determination of microbiological and chemical risk and its mitigation drives our research program. In addition, FDA is doing extensive research on the detection, characterization, and behavior of foodborne pathogens, microbial genetics, and molecular virology. These efforts are vital in our attempt to develop risk assessment models for pathogens and intervention strategies to reduce the public health risk that these pathogens present. FDA's research in the area of chemical contaminants focuses on the development of detection methods and toxicology studies.

Collaborative research efforts further strengthen the scientific basis for our food safety programs. For example, for the past decade, FDA has worked closely with USDA's Agricultural Research Service (ARS) and Cooperative State Research, Education, and Extension Service (CSREES) to coordinate and mutually support our respective research efforts related to produce safety. This relationship allows FDA to augment its research resources and gain access to facilities and expertise we do not have. In this spirit, we collaborated with ARS and CSREES to analyze water samples from the Salinas watershed for *E. coli* O157:H7, and to relate the location of bacteria to geographical, seasonal, or rainfall variation. An extension of this research will look for sources of *E. coli* O157:H7 in California's Salinas Valley. Information obtained from this study will be used to inform produce growers about strategies to prevent pre-harvest microbial contamination.

We also strengthen the scientific basis for our program by collaborating and learning with others, such as participating in many scientific and technical meetings on food safety. In February, we participated in a forum sponsored by the Western Institute for Food Safety and Security to share information on assessing industry approaches to address the safety of lettuce and leafy greens on the farm and at packing, cooling, and processing facilities. Also in February, the FDA-affiliated Joint Institute for Food Safety and Applied Nutrition and the University of Florida sponsored a workshop to improve understanding of how tomatoes become contaminated with *Salmonella* and other pathogens. This month on May 30th and 31st, FDA, the National Center for Food Safety and Technology, and the University of Georgia's Center for Food Safety will co-sponsor a workshop on microbial testing to reach a consensus on the role of microbial testing in ensuring the safety of produce.

In response to the recent outbreaks, FDA recently held two public hearings concerning the safety of fresh produce. The purpose of these hearings was for FDA to share information about recent outbreaks of foodborne illness related to fresh produce and to solicit comments, data, and additional scientific information on this issue. We are soliciting input from all our stakeholders on ways to improve the safety of fresh produce. The administrative record will remain open until June 13, 2007.

Enhancing Effective Partnerships

To succeed in our science-based efforts to promote food safety, we need to enhance our collaborations with stakeholders interested in food safety, particularly with respect to fresh produce. Fresh produce is produced on tens of thousands of farms, and contamination at one step in the growing and processing chain can be amplified throughout the following steps. FDA has worked with the public and private sector to encourage industry to follow the recommendations and standards contained in FDA guidances. After enlisting the help of the scientific community and industry, FDA published the "Guide to Minimize Microbial Food Safety Hazards for Fresh Fruits and Vegetables." This guide, published in 1998, recommends good agricultural practices and good manufacturing practices that growers, packers, and shippers can take to address common risk factors in their operations. FDA and USDA issued the guide in several languages and have conducted significant outreach, both domestically and internationally, to encourage its implementation. In addition, FDA has assisted industry in developing a number of commodity-specific food safety guidelines for the commodities most often associated with foodborne illness outbreaks. These include guidelines for lettuce and leafy greens, melons, and tomatoes. We are working with industry on similar guidelines for herbs and green onions.

The example of fresh sprouts illustrates how successful these efforts can be. In 1999, there were 390 reported illnesses associated with eating contaminated fresh sprouts. FDA published two guidance documents for sprouts that year. In 2004, only 33 illnesses were reported associated with fresh sprouts and, in 2005 and 2006, there were none. We believe that the decline in sprout-associated illnesses was in large part due to the industry's adherence to the recommendations FDA provided in those guidance documents through our outreach and inspection efforts.

FDA's efforts in this area are ongoing. In March, FDA issued a draft final version of its "Guide to Minimize Microbial Food Safety Hazards of Fresh-cut Fruits and Vegetables" (the Fresh-cut Guide). This guidance is intended for all fresh-cut produce firms, including, among others, those that process fresh-cut spinach and lettuce/leafy greens, to enhance the safety of fresh-cut produce by minimizing the microbial food safety hazards. In addition, based on FDA's efforts, the Codex Alimentarius Commission, the international food safety standards body, has asked the Food and Agriculture Organization/World Health Organization (FAO/WHO) for an expert consultation on the microbiological safety of fresh produce to support the development of commodity-specific annexes to the hygienic code. FAO/WHO just announced that this consultation will occur during 2007 and early 2008.

In August 2006, FDA launched its "Lettuce and Leafy Greens Initiative," which assesses practices and conditions at select farms and facilities in California, in collaboration with California's Department of Health Services and its Department of Food and Agriculture. We will continue to work with Federal, state, local and international food safety partners and with industry to develop guidance, conduct research, develop educational outreach materials, and initiate other commodity—or region-specific programs to enhance the safety of fresh produce.

The close collaboration between Federal and state food safety officials in response to the *E. coli* O157:H7 outbreak associated with fresh spinach is a good example of the effective working relationships we enjoy with our food safety partners. On March 23, FDA and California's Department of Health Services (CDHS) released a joint report on an extensive investigation into the causes of the *E.coli* O157:H7 outbreak last fall that was associated with contaminated Dole brand baby spinach and resulted in 204 confirmed illnesses and three deaths. The inquiry was conducted by the California Food Emergency Response Team (CalFERT), a team of experts from FDA's district office in San Francisco and CDHS. Potential environmental risk factors for *E.coli* O157:H7 contamination identified in the report included the presence of wild pigs, the proximity of irrigation wells used to grow produce for ready-to-eat packaging, and surface waterways exposed to feces from cattle and wildlife. Because the contamination occurred before the start of the investigation, and because of the many ways that *E.coli* O157:H7 can be transferred — including animals, humans, and water — the precise means by which the bacteria spread to the spinach remain unknown.

Another important example of a food safety partnership we continue to enhance is the FDA/USDA Food Emergency Response Network (FERN). FERN is a network of Federal, state, and local laboratories capable of testing food samples for microbiological, chemical, and radiological threat agents. This partnership provides essential analytical expertise and surge capacity in case of emergencies. The number of participating laboratories has increased to 134 laboratories in FY 2007, compared to 30 participating laboratories in March 2004 (near FERN's inception). The FERN network proved to be a critical asset in the *E. coli* O157:H7 outbreak associated with fresh spinach. FERN analysts worked closely with CDC's Laboratory Response Network personnel to harmonize and approve a modified FERN method for detecting *E. coli* O157:H7 in spinach. This method allowed for substantially improved testing of spinach samples as it allowed for the detection of *E. coli* O157:H7 at lower levels.

Improving Risk-Based Targeting of Inspection Resources

FDA is significantly improving its ability to target its inspection resources at the greatest risks to public health. However, inspections cannot and will not identify every potential food safety problem. Improving the processes and operations of all participants in the food production and distribution process offers the greatest protection for American consumers, and inspections are only one component of this activity. To make best use of available resources, FDA uses a targeted, risk-based approach to inspections. For example, we inspect almost all high-risk food establishments on an annual basis. FDA's electronic screening system for imported food shipments also uses a risk-based approach and allows us to focus our physical inspection resources on what appear to be higher risk shipments. In 2006, FDA performed over 89,000 security reviews on shipments of food and animal feed offered for import and flagged as high risk by our screening system. In addition, we performed over 94,000 field exams on imported food products selected for physical examination by our risk-based screening criteria. To enhance our ability to target our inspections, we are working toward a system that will include information from all points on the import life cycle, from production to consumption, in order to better protect consumers.

CONCLUSION

FDA is working hard to ensure the safety of food, in collaboration with its Federal, state, local, and international food safety partners, and with industry, consumers, and academia. As a result of this effective collaboration, the American food supply continues to be among the safest in the world. We have made progress, but the recent incidents of contaminated food and animal feed demonstrate the challenges we face and the need to enhance our efforts. As I noted at the beginning of my testimony, we are working on a new strategy for food safety and food defense that will address changes in the global food safety and food defense system, identify our most critical needs, and serve as a framework for addressing the challenges we face. We will continue to strive to reduce the incidence of foodborne illness to the lowest level possible.

Thank you for the opportunity to discuss FDA's continuing efforts to improve the safety of fresh produce. I would be happy to answer any questions.

In: Food Safety, Fresh Produce and FDA Oversight
Editor: Russell P. Barton
ISBN: 978-1-60692-788-5

Chapter 7

TESTIMONY BEFORE THE COMMITTEE ON AGRICULTURE SUBCOMMITTEE ON HORTICULTURE AND ORGANIC AGRICULTURE, UNITED STATES HOUSE OF REPRESENTATIVES

Lonnie J. King

INTRODUCTION

Good afternoon, Chairman Cardoza and Members of the Subcommittee. I am Dr. Lonnie King, Director of the National Center for Zoonotic, Vector-borne, and Enteric Diseases, at the Centers for Disease Control and Prevention. Thank you for the invitation to address the Subcommittee on CDC's activities related to the prevention of foodborne disease and CDC's role in the response to the current outbreak of *Salmonella* Saintpaul infections associated with fresh produce. First, let me offer my sympathies to all the families who have been adversely affected by this outbreak. Second, I understand the frustration of many in the food producing and serving industries, who work very hard to produce and serve safe produce. This investigation has been especially difficult and prolonged. We have faced many challenges with this particular foodborne outbreak. I will discuss these challenges in more detail after describing the CDC's response to the *Salmonella* Saintpaul outbreak.

BACKGROUND

Foodborne disease presents a continuing challenge to public health. CDC estimates that approximately 76 million U.S. residents get sick, 325,000 are hospitalized, and 5,000 die each year from foodborne illness. Overall, foodborne diseases appear to cause more illnesses but fewer deaths than previously estimated in the 1980's. More than 250 different foodborne illnesses have been described. Most are caused by a variety of bacteria, viruses, and parasites. Some foodborne illnesses are caused by toxins or chemicals.

As an agency within the Department of Health and Human Services (HHS), CDC leads federal efforts to gather data on foodborne illnesses, investigate foodborne illnesses and outbreaks, and monitor the effectiveness of prevention and control efforts. CDC is not a food safety regulatory agency but works closely with the food safety regulatory agencies, in particular with HHS's Food and Drug Administration (FDA) and the Food Safety and Inspection Service within the United States Department of Agriculture (USDA). CDC also plays a key role in building state and local health department epidemiology, laboratory, and environmental health capacity to support foodborne disease surveillance and outbreak response. Notably, CDC data can be used to help document the effectiveness of regulatory interventions.

Much of what CDC does depends on critical partnerships with state and local public health departments who collect surveillance data and investigate most outbreaks themselves. CDC has worked with the Association of Public Health Laboratories (APHL) and the Council of State and Territorial Epidemiologists (CSTE) to strengthen networks for foodborne disease surveillance. For example, PulseNet, the national network for molecular subtyping of foodborne bacteria coordinated by CDC, empowers every state health laboratory to test strains of bacteria from sick persons in that state, and to compare them with DNA "fingerprint" patterns in the national database at CDC. This has greatly improved the ability to detect clusters of illness that may be related, even if they are dispersed across multiple states.

OutbreakNet is the group of public health officials at State health departments and CDC who regularly investigate foodborne outbreaks. The OutbreakNet team at CDC coordinates the investigation of the large, multistate clusters and works with the foodborne disease epidemiologists in each state to evaluate clusters that PulseNet detects. The OutbreakNet team at CDC also manages the electronic Foodborne Outbreak Reporting System (eFORS). Established in 2001, eFORS is a web-based outbreak surveillance system through which state and local health departments voluntarily submit completed reports of foodborne disease outbreak investigations to CDC.

CDC's Environmental Health Specialists Network (EHS-Net), a collaborative effort with FDA and nine states, assists state health departments in their efforts to improve the practice of environmental health service programs; participants assess policies and practices of retail foodservice establishments that could lead to or prevent foodborne outbreaks. FoodNet is a network that is a collaborative effort among CDC, 10 states who participate in CDC's Emerging Infections Program, the Department of Agriculture's Food Safety and Inspection Service (UDSA/FSIS), and FDA; it provides the most accurate surveillance data for determining the burden of infections, conducts scientific studies to better understand the sources for the many illnesses that occur outside the outbreak setting, and monitors trends in infections as new control measures are instituted. We have PulseNet to detect possible outbreaks, OutbreakNet to investigate and report them, and FoodNet to track general trends and define where more effective prevention strategies are needed.

CDC also works with a broad range of other partners to improve capacity and knowledge regarding foodborne disease control and prevention. In collaboration with the National Environmental Health Association (NEHA), CDC conducts team training programs for local and state health department officials including specialists in environmental health, laboratory, and epidemiology. CDC works with the World Health Organization (WHO) and a variety of other international partners to conduct similar training programs in other countries through the WHO Global Salmonella Surveillance program. CDC supports the Council to Improve Foodborne Outbreak Response (CIFOR) which was created to help develop model programs and processes that will facilitate the investigation and control of foodborne disease outbreaks. CSTE and the National Association of County and City Health Officials (NACCHO) are co-chairing CIFOR, and it includes representatives from CDC, FDA, USDA, APHL, NEHA, the Association of State and Territorial Health Officials, and the Association of Food and Drug Officials.

Salmonella

Salmonella is a group of bacteria that is widespread in the intestines of birds, reptiles, and mammals. *Salmonella* bacteria have been known for over 100 years to cause human illness. *Salmonella* is the second most common bacterial cause of foodborne diseases in the country, causing 15 reported laboratory-confirmed infections per 100,000 population in 2007, as measured in FoodNet. There are many different kinds, or serotypes, of *Salmonella* bacteria. Serotyping is a classification system based on differences in structures on the surfaces of bacteria or other disease-causing agents. Serotyping divides *Salmonella* into more than 2500 different serotypes, some common and some rare. For example, during 1996-2006, *Salmonella* serotype Typhimurium and *Salmonella* serotype Enteritidis typically caused 41% of reported *Salmonella* illnesses each year in the United States. *Salmonella* serotype Saintpaul is relatively uncommon, causing only 1% (about 400) of all reported laboratory-confirmed *Salmonella* infections each year. Each serotype can be further sub-divided into many more subtypes based on their DNA.

Salmonella infections have often been associated with meat, poultry, eggs, and raw milk; these products are derived from animals that can carry *Salmonella*. *Salmonella* has also been associated with fresh produce and other plant-derived foods. Fresh produce can be an important source of other types of foodborne infections as well; for example, *Escherichia coli* 0157, another bacterial agent, caused a large outbreak of illness linked to spinach in 2006. *Salmonella*, like other pathogens that are commonly foodborne, can also be transmitted in other ways, such as from contact with reptiles or other animals or between children at a child care center.

Many foodborne infections, including *Salmonella*, occur in persons without obvious connections to each other. These are called sporadic cases; determining the source of a single sporadic case can be very difficult. Cases of similar infections can also occur as a group or "cluster." Epidemiological investigation of clusters of possibly related cases permits public health officials to determine if the cases were connected and, specifically, if they were linked to food. A cluster of foodborne illnesses is considered an outbreak if an investigation demonstrates that two or more infections caused by the same agent are linked to the same food.

In general, for a foodborne illness to be recognized by the public health surveillance system, a patient must seek medical attention, the physician must decide to order diagnostic tests, and the laboratory must conduct the test using the appropriate procedures and report the results to a health department. Many ill people do not seek medical attention, and of those who do, many are not tested. Therefore, many cases of foodborne illness are neither diagnosed nor reported. For example, *Salmonella* infection has been estimated to cause about

1.4 million foodborne illnesses annually, however, only about 40,000 laboratory-confirmed cases of *Salmonella* are reported to CDC each year.

Regular reporting about detection of *Salmonella* serotypes from ill persons is critical in determining whether a change in incidence has occurred signaling a possible outbreak. Each serotype can be further divided by DNA analysis into subtypes. The subtypes are distinguished by different DNA fingerprint patterns. The fingerprint pattern is determined with a test known as pulsed-field gel electrophoresis (PFGE). PFGE is a very good method for discriminating between epidemiologically unrelated isolates of this serotype. Public health laboratories determine the serotype and PFGE patterns for *Salmonella* strains and share the patterns through PulseNet. PulseNet plays a vital role in surveillance for and investigation of widely dispersed foodborne illness outbreaks that were previously difficult to detect. The laboratories participating in PulseNet are in state health departments, some local health departments, USDA, and FDA. When a clinical laboratory detects *Salmonella* from an ill person, a sample is sent to a State or local PulseNet laboratory where it is serotyped and DNA fingerprinted. The laboratory compares the fingerprint pattern to that of other *Salmonella* strains from people in that area and uploads the pattern electronically to the national PulseNet database maintained at CDC, where it can be compared with the patterns from all over the country. This gives us the capability to detect an unusual number of *Salmonella* cases with the same pattern in a single area or in multiple states. The system can identify patterns even if the affected persons live far apart, which is important given the widespread U.S. food distribution systems. The pattern causing the current outbreak is usually quite uncommon, and was identified only 25 times in 2007, among the 400 *Salmonella* Saintpaul infections that were reported.

It is important to recognize there is an inherent delay between when a person becomes ill with *Salmonella* infection and when the results of testing are reported to PulseNet. In the current *Salmonella* Saintpaul outbreak, the median number of days between when the illness began and when the fingerprint pattern was reported to PulseNet has been 16 days. It takes time for a person to become ill, seek medical care, submit a sample for testing; it then takes time for the clinical laboratory to detect *Salmonella* and send the strain to the public health laboratory; it then takes time for the public health laboratory to perform serotyping and DNA fingerprinting.

The Salmonella Saintpaul Outbreak

On May 22, 2008, the New Mexico Department of Health contacted CDC to report that they were investigating illness in 4 persons with *Salmonella* Saintpaul strains that had the same DNA fingerprint pattern, and that *Salmonella* strains from 15 more persons were still being characterized. The DNA fingerprint determined by PFGE was rare. It usually occurs no more than 2-3 times a month in the whole United States, so 4 or more in one location was unexpectedly high. New Mexico posted the information about the unusual number of *Salmonella* Saintpaul cases to the PulseNet web board on May 22, so that all state laboratories could quickly compare the DNA fingerprint pattern with that of their own strains, and CDC requested that states report any strains that matched the DNA fingerprint pattern. That next day, Texas and Colorado reported cases with this PFGE pattern, and investigators in the New Mexico Department of Health, the Navajo Nation, the Indian Health Service, the Texas

Department of State Health Services, and CDC began a multistate investigation. Daily multistate conference calls began and continued through July, with states being added to the calls as their cases were identified. The investigation was initially coordinated by the New Mexico State Health Department, because most identified cases were in that state. On June 3, after more states in different regions of the country reported cases, CDC assumed this role of the investigation.

The initial steps in an epidemiological investigation are to collect information from which hypotheses can be generated about the possible source of the outbreak. As cases with the same DNA fingerprint pattern were identified, epidemiologists interviewed patients to determine what specific foods or other exposures they may have had in common. The New Mexico Department of Health, Texas Department of State Health Services, and the Indian Health Service conducted hypothesis-generating interviews from mid- to late- May among 19 ill persons from whom *Salmonella* Saintpaul with the DNA fingerprint matching the outbreak strain had been isolated during May 2008. These interviews collected information about possible sources of infection, including attendance at gatherings, travel, daycare contact, contact with reptiles and/or other household pets, contact with farm animals, sources of drinking water, history of swimming, eating at restaurants, and specific food consumption history for approximately 200 food items; the interviews also included open-ended questions about what ill persons had eaten, meal by meal, in the days before they became ill. The preliminary results of this first series of interviews indicated raw tomatoes were the most commonly consumed food item (reported by 84% of ill persons) leading to the hypothesis that they were a possible source of the illnesses. CDC informally advised the FDA on May 26 of the hypothesis of a possible association between ill persons and the consumption of raw tomatoes.

In the next steps of the investigation, analytic epidemiologic studies were conducted to test the hypotheses generated by case finding. These studies compare the frequency with which ill persons report exposure to a particular food item to the frequency with which healthy persons (or controls) report that exposure. If the ill group is more likely than the well group to report exposure to a particular food, a statistical test can show how likely this finding would have occurred by chance alone. Additional information about the likelihood of that particular food actually being contaminated, the biological plausibility of it causing the illnesses, the fit of the cases with the distribution of the food, and other factors may enter into the professional judgment of whether the food with a statistically significant association with cases is likely to explain the outbreak. Preliminary findings from these types of studies guided subsequent next steps of the investigation while additional statistical analyses are being conducted on the data gathered. It is important to keep in mind at this stage of investigation, as analyses are conducted and interpreted, that initial findings and hypotheses may change. As is common in outbreak investigations but especially true for foodborne outbreaks, the process of case finding, hypothesis generating, and hypothesis testing is an iterative process; each step informs subsequent steps and often leads to new investigative avenues.

In the next phase of the investigation, in late May, the New Mexico Department of Health, the Texas Department of State Health Services, and the Indian Health Service, in consultation with CDC, conducted a multi-state case-control study. The data from the earlier 19 hypothesis generating interviews were used to identify which foods were most frequently consumed by the ill people. The questionnaire used in this case-control study included the 14 foods[1] reported by half or more of the ill people in the hypothesis-generating interviews.

These questionnaires were administered to approximately 150 people. By May 31, preliminary results of the case-control study demonstrated that illness was significantly associated with consumption of raw tomatoes (88% of cases consumed raw tomatoes compared with 64% of the controls, a very strong statistical difference). FDA was formally notified of this significant association between tomatoes and infection. Statistical analysis of these data showed that illness was associated with consumption of raw tomatoes independent of consumption of tomatoes in salsa, guacamole, or pico de gallo.

The next step in the investigation was to trace the implicated food back to its sources, looking for points where contamination might have occurred, and to determine if there is a single farm, processing location, or other point in distribution system that could explain all the illnesses providing additional evidence supporting the food item as a cause of the outbreak. Tracing the implicated food back from consumption through preparation, to distributors, and source can also help determine how the contamination occurred, stop distribution of the contaminated product, and prevent further outbreaks from occurring. On May 31, 2008, FDA decided to initiate investigations attempting to trace the tomatoes reported to have been eaten by ill persons back to their sources. Tracebacks began on June 1, 2008. Throughout the investigation there has been ongoing communication between CDC and FDA regarding these traceback investigations.

On June 4, CDC received the first report of a possible restaurant cluster. Four cases in Illinois appeared to be related to exposure to a single restaurant. Such clusters were otherwise absent in the early part of the outbreak. The outbreak continued and expanded. Over the next few weeks, hundreds more cases were reported from an increasing number of states. The average number of persons who became ill between May 20 and June 10 was 33 per day. New information emerged as each case was reported and interviewed by local or state health department authorities.

On June 16, CDC learned about the first recognized large cluster linked to a single restaurant, approximately 30 illnesses, in Texas. Between June 18 and June 20, Texas reported an additional 134 cases. This surge in the number of cases from Texas highlighted the geographic concentration in the Southwest and in Native American and Hispanic persons, which did not have a clear explanation. This information, along with the strong association between illness and consumption of Mexican-style foods in restaurants coming from continued analysis of the case-control studies, and the apparent continuation of the outbreak after the alert regarding tomatoes, led to the hypothesis that a food item commonly consumed with tomatoes could be causing illnesses. Epidemiologists decided to focus the investigations on the recently identified clusters and to conduct a case-control study of persons nationwide who became ill in June. CDC offered assistance to the Texas Department of State Health Services; a CDC Epi-Aid team arrived in Texas on June 19.

By July 7, 32 clusters of *Salmonella* Saintpaul infections with the PFGE pattern of the outbreak strain had been identified in 13 states and the District of Columbia. Twenty-six were associated with Mexican-style restaurants. Most clusters had fewer than 5 ill persons. Three clusters had more than 10 ill persons, and analytic studies have been conducted on these. In one of these larger restaurant clusters, illnesses were linked to consumption of an item containing fresh tomatoes and fresh jalapeño peppers. In the other two, illnesses were linked to an item containing fresh jalapeño peppers but neither raw tomatoes, nor fresh cilantro. Among the 22 smaller clusters with data on the presence of food items in the venue, four did not serve jalapeño peppers. Together, these investigations indicated that jalapeño peppers caused some illnesses, but did not appear to explain all illnesses. Raw tomatoes, fresh serrano peppers, and fresh cilantro also remained under investigation. We were strongly considering the probability that more than one food item caused illness.

CDC and state and local health departments conducted a second case-control study to investigate the possibilities that illness was related to consuming foods in Mexican-style restaurants, and that illness was associated with consuming, in a restaurant, event, or home, a range of produce items that are often served with tomatoes, including freshly made salsa, fresh jalapeño peppers, and fresh cilantro. This was a large multistate study, with over 400 interviews, with 141 interviews from persons who had become ill on or after June 1 and 281 interviews from healthy controls available for preliminary analysis. The study showed that illness was strongly associated with eating at a Mexican-style restaurant. In a preliminary statistical analysis that considered the entire dataset, consumption of fresh tomatoes, jalapeño peppers, and cilantro were each shown to be risk factors in subgroups but no single suspect exposure statistically dominated the others in explaining all cases. Thus, this study indicated a strong link to fresh produce items used in Mexican cuisine but did not point clearly to one specific item.

As new restaurant-associated clusters were reported, CDC and state health departments investigated them aggressively. By July 16, CDC investigators were assisting state and local health officials in field investigations of restaurant clusters in North Carolina, Missouri, Texas, and New York City. In addition, another CDC team was investigating illnesses in New Mexico.

As the epidemiological investigation expanded, the FDA also expanded their traceback and sampling efforts. FDA began their tracebacks on peppers identified by the outbreak investigations conducted by the states and CDC. CDC sent two medical epidemiologists to FDA to directly participate in analyzing findings from the tracebacks and connect them with the CDC epidemiologic data. On July 21, the FDA announced that they had isolated the outbreak strain of *Salmonella* Saintpaul from a sample of jalapeño peppers. The epidemiologic data from a Texas cluster of ill persons led to this specific traceback investigation. In most *Salmonella* outbreaks that are linked to a particular food, however, *Salmonella* is never detected in the food. Detection of *Salmonella* in a food item that was implicated in an epidemiologic study provides strong evidence that this food item caused illnesses, though it does not exclude other foods as possible causes of illness.

Throughout the investigative process, to ensure that information was disseminated to the public as accurately and quickly as possible about health threats and other information related to this outbreak, CDC and FDA coordinated their communication strategies and messages and discussed these strategies in daily calls with state health officials. We balance the rapid release of information on sources of illness against the potential negative consequences to consumers, food growers, producers, and industry. Continued collaborations and communications between federal agencies, state and local health departments, and all relevant stakeholders are essential.

Challenges Confronting the Outbreak Investigation

Every outbreak response is a challenge for everyone involved. This outbreak was particularly challenging in a number of ways. As already mentioned, it takes time for a case to be reported to public health authorities and then investigated. For half the cases in this outbreak, it took more than 16 days from when the person became ill to the when the DNA fingerprint of their *Salmonella* was added to the PulseNet database. The resulting delay sometimes prevented interviews from occurring while memories were still fresh. The precision of interviews by epidemiologists depend on the observations and memories of people about what they ate and what ingredients the dishes contained. People often have difficulty remembering exactly what foods they ate, and remembering specific ingredients in those foods is even more difficult, especially if the dish was prepared by someone else, or eaten in a restaurant. Another challenge has been that the foods in question are often eaten together – many salsa, guacamole, and pico de gallo recipes contain tomatoes, jalapeño peppers, and cilantro, so exposure to one item often means exposure to all three. When food items are mixed together and consumed in the same dish, all the items may be statistically linked to illness. In that case, it can be difficult or impossible to separate out the risk from individual foods without additional information such as microbiological culture or traceback of the foods. Although laboratory testing of foods might help identify the source of an outbreak, perishable foods that were consumed by ill persons were often not available to test. This is in contrast to outbreaks from frozen or processed foods which may still be present in someone's freezer or pantry weeks later. Finally, the traceback of fresh produce, such as tomatoes, through the supply chain can be very difficult and labor intensive. Doctor Acheson will be able to say more about this.

Status of Investigation

As of July 27, 1304 persons infected with *Salmonella* Saintpaul with the same fingerprint have been identified in 43 states, the District of Columbia, and Canada. At least 252 persons were hospitalized. Two deaths were possibly linked to the outbreak: A man in his eighties who died in Texas from cardiopulmonary failure had an infection with the outbreak strain at the time of his death. A man in his sixties who died in Texas from cancer had an infection with the outbreak strain at the time of his death.

Three larger clusters were intensively investigated as of July 7. In one, illnesses were linked to consumption of an item containing fresh tomatoes and fresh jalapeño peppers. In the other two, illnesses were linked to an item containing fresh jalapeño peppers and no other of the suspect items. Since then, detailed investigations of three other clusters indicate that jalapeño peppers do not explain all illnesses. In two of these more recent investigations, illnesses were linked to an item containing fresh serrano peppers and tomatoes, but not jalapeño peppers. In a third, illnesses were linked to an item that contained fresh jalapeños and tomatoes. Other clusters are under active investigation. At present, the information indicates that jalapeño peppers and serrano peppers grown, harvested, or packed in Mexico are the cause of some clusters and could be a major food vehicle for the outbreak. The U.S. Food and Drug Administration is advising consumers that jalapeno and serrano peppers grown in the United States are not connected with the current *Salmonella* Saintpaul outbreak and consumers may feel free to eat them without concern of illness. By themselves, tomatoes cannot explain the entire outbreak, nor do jalapeño peppers explain all the clusters. The outbreak appears to be ongoing, but with fewer new illnesses each day. New, very active field investigations by CDC in collaboration with State and local health departments, and FDA tracebacks on jalapeño peppers and tomatoes are providing new information almost daily. It appears likely that more than one food vehicle is involved. Although rare, more than one food has been implicated in foodborne outbreaks in the past, as observed in a group of 1998 outbreaks traced to imported parsley and cilantro from a single farm.

Conclusion

The current outbreak investigation of *Salmonella* Saintpaul is the largest foodborne outbreak in the United States in the past decade. The investigation has been especially complex, difficult, and prolonged. The outbreak appears to be slowing, but we are not able to say with confidence that the outbreak is over because of the reporting delay. The event illustrates how a large and widespread outbreak can occur, appearing first as individual cases, then as small clusters, and finally with large numbers of persons becoming ill if a widely consumed food is contaminated. It also illustrates the importance of existing public health networks: the laboratories performing PulseNet fingerprinting; the epidemiologists conducting the investigation; the environmental health aspects of the outbreak; the multi-disciplinary approach to the investigation; and the close communication and collaboration among local, state, and federal officials.

CDC is prepared to continue working with regulatory authorities, state and local partners, food and environmental microbiologist scientists, and the food industry to find long-term solutions to this challenging problem.

Thank you again for the invitation to testify before you today. I will be happy to answer any questions you may have.

End Notes

[1] Food items examined included tomatoes, eggs, ice cream, potatoes, milk, tortillas, cold breakfast cereal, raw onions, salsa, avocado, guacamole, ground beef, chicken, and lettuce.

CHAPTER SOURCES

The following chapters have been previously published:

Chapter 1 – This is an edited, reformatted and augmented version of a United States Government Accountability Office Report to Congressional Requesters, GAO-08-1047, dated September 2008.

Chapter 2 – This is an edited, reformatted and augmented version of a Congressional Research Service publication, Order Code RS22600, updated June 9, 2008.

Chapter 3 – This is an edited, reformatted and augmented version of a Congressional Research Service publication, Order Code RS22779, updated February 6, 2008.

Chapter 4 – This is an edited, reformatted and augmented version of a Congressional Research Service publication, Order Code RS22946, updated September 8, 2008.

Chapter 5 - This is an edited, reformatted and augmented version of remarks delivered as Statement of David Acheson, M.D., F.R.C.P. before the Committee on Agriculture Subcommittee on Horticulture and Organic Agriculture United States House of Representatives, dated May 15, 2007.

Chapter 6 - This is an edited, reformatted and augmented version of remarks delivered as Statement of David Acheson, M.D., F.R.C.P. before the Committee on Agriculture Subcommittee on Horticulture and Organic Agriculture United States House of Representatives, dated July 30, 2008.

Chapter 7 - This is an edited, reformatted and augmented version of remarks delivered as Testimony of Lonnie J. King, D.V.M. before the Committee on Agriculture Subcommittee on Horticulture and Organic Agriculture United States House of Representatives, dated Wednesday, July 30, 2008.

INDEX

A

B

C

D

E

F

G

H

I

J

L

M

N

O

P

Q

R

S

T

U

V

W